INDUSTRIAL ECOLOGY AND SUSTAINABILITY

A Textbook for Students

INDUSTRIAL ECOLOGY AND SUSTAINABILITY

A Textbook for Students

Thomas E Graedel
Yale University, USA

Matthew J Eckelman
Northeastern University, USA

World Scientific

NEW JERSEY • LONDON • SINGAPORE • BEIJING • SHANGHAI • TAIPEI • CHENNAI

Published by

World Scientific Publishing Co. Pte. Ltd.

5 Toh Tuck Link, Singapore 596224

USA office: 27 Warren Street, Suite 401-402, Hackensack, NJ 07601

UK office: 57 Shelton Street, Covent Garden, London WC2H 9HE

Library of Congress Cataloging-in-Publication Data

Names: Graedel, T. E., author. | Eckelman, Matthew J., author.

Title: Industrial ecology and sustainability : a textbook for students / Thomas E. Graedel, Yale University, USA, Matthew J. Eckelman, Northeastern University, USA.

Description: Hackensack, NJ : World Scientific Publishing Co. Pte. Ltd., [2025] | Includes bibliographical references and index.

Identifiers: LCCN 2024032525 | ISBN 9789811296147 (hardcover) | ISBN 9789811297564 (paperback) | ISBN 9789811296154 (ebook for institutions) | ISBN 9789811296161 (ebook for individuals)

Subjects: LCSH: Industrial ecology. | Sustainable development.

Classification: LCC TS161 .G7424 2025 | DDC 363.17--dc23/eng/20241118

LC record available at https://lccn.loc.gov/2024032525

British Library Cataloguing-in-Publication Data

A catalogue record for this book is available from the British Library.

For any available supplementary material, please visit
https://www.worldscientific.com/worldscibooks/10.1142/13929#t=suppl

Desk Editors: Kannan Krishnan/Amanda Yun

Typeset by Stallion Press
Email: enquiries@stallionpress.com

To Susannah and Gwyneth
Partners and Inspirers

Preface

This book is derived from our 2023 comprehensive overview of industrial ecology and its applications, titled simply Industrial Ecology and Sustainability. From that work we have extracted and rewritten selected chapters that introduce the specialty of industrial ecology and present and discuss the methodological approaches employed widely in the field of study. We discuss the "grand challenges" imposed upon modern society by global population growth, increased demand for a wide variety of natural resources, and the growing challenges of enabling sustainable societies in the midst of rapidly evolving climate transformations.

In 2007 Johnathon Porritt, a British environmental leader for half a century, issued a call to engineering educators and to engineers in training: "It is monumentally stupid to let graduates leave the university in a state of sustainability illiteracy". Almost two decades later, the situation is little changed for engineering students, for natural science students and, indeed, for almost all students.

The field of study in the present volume thus addresses major challenges in technology and society, and thus finds itself at the center of a potential sustainability transition for planet Earth. We hope that this book will help provide tools and insights to assist industrial ecology educators and practitioners as they rise to meet the challenges ahead.

Thomas E. Graedel
New Haven, CT, USA

Matthew J. Eckelman
Boston, MA, USA

About the Authors

Thomas E. Graedel joined Yale University in 1997 after 27 years at AT&T Bell Laboratories. He is currently a Professor Emeritus and Senior Research Scientist at Yale. One of the founders of the field of industrial ecology, he co-authored the first textbook in that specialty and has published extensively and lectured widely on industrial ecology's implementation and implications. His characterizations of the cycles of industrially used metals have explored aspects of resource availability, potential environmental impacts, opportunities for recycling and reuse, materials criticality, and resources policy. He was the inaugural President of the International Society for Industrial Ecology from 2002 to 2004 and the winner of the ISIE Society Prize for excellence in industrial ecology research in 2007. He has served three terms on the United Nations International Resource Panel and was elected to the US National Academy of Engineering in 2002.

Matthew J. Eckelman is an Associate Professor of civil and environmental engineering at Northeastern University and an Adjunct Associate Professor at the Yale School of Public Health. His research laboratory builds processed-based emissions models and lifecycle sustainability assessment tools. He has served as a Board Member and Treasurer of the International Society for Industrial Ecology and received the Laudise Medal for research in industrial ecology in 2013. For the past 15 years, he has also served as CTO of the green engineering firm Sustainability A to Z, LLC (currently based in Connecticut, USA) providing environmental consulting services to Fortune 500 companies, industry associations, and public agencies. He is a member of the Lancet Countdown on Health and Climate Change and the National Academy of Medicine Decarbonization Action Collaborative. Dr Eckelman worked previously for the Massachusetts Executive Office of Environmental Affairs and received a PhD in Chemical and Environmental Engineering from Yale University in 2010.

Contents

Introducing Industrial Ecology

Chapter 1

Humanity, Technology, and Sustainability

Chapter Concepts

- Common-pool resources are threatened by individual self-serving actions unless some form of community action intervenes.
- Technological innovations can drastically change the types and quantities of energy and materials that humanity uses to provide services.
- The "master equation" of industrial ecology defines environmental impact as a function of population, affluence, and technology.

1.1 Our Technological World

Planet Earth has never been static, even well before the arrival of humans. Glacial action, weathering, erosion, sediment transport, and other natural processes have moved large amounts of soil and sediment for eons. Since human populations grew and began to harness large quantities of resources for their own purposes, however, the relative importance of natural processes compared to those of human actions has dramatically changed. Estimates suggest that human agricultural land-altering activity exceeded natural processes several millennia ago, but since the Industrial Revolution around the latter part of the 17th century human action has accelerated rapidly. A dramatic example of the more recent human footprint is shown in Figure 1.1: that as of about 2020 the amount of human-made biomass

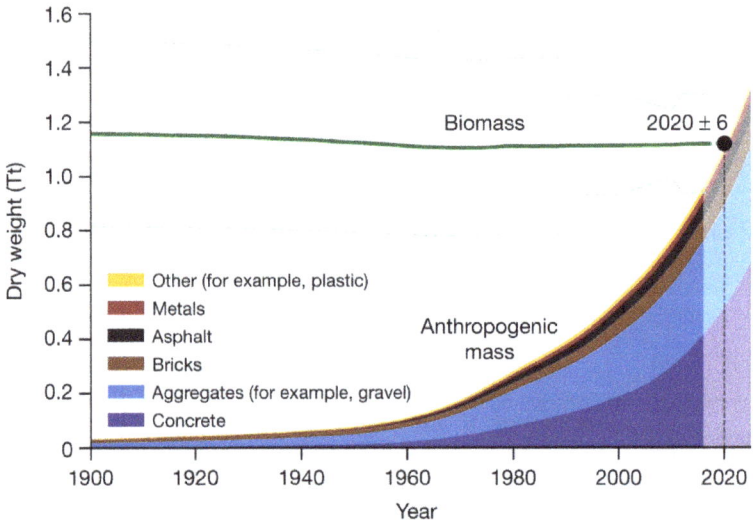

Figure 1.1. Biomass and anthropogenic mass estimates since 1900 (dry-mass basis); the dashed green lines represent one standard deviation in the biomass estimate (Elhacham *et al.*, 2020).

(concrete, bricks, gravel, asphalt, metals, plastics, etc.) had equaled or exceeded all of Earth's biomass and is continuing to increase rapidly. It has become clear that on almost every part of our planet, technologically-enabled human action has become dominant over the forces of geology and ecology. Images like those of Figure 1.1 provide much of the impetus for referring to the current time period as the Anthropocene: the epoch of human dominance of the planet.

1.2 The Tragedy of the Commons

In 1968, Garrett Hardin, an ecology professor at the University of California, Santa Barbara, published an article titled "The Tragedy of the Commons" that has remained relevant, perhaps even more so with each passing year. Hardin argued that a society that permitted individuals to engage in activities that adversely influenced common resources was eventually doomed to failure. Hardin cited as an example a community pasture area, used by any local herdsman who wishes to. Each herdsman, seeking to maximize financial well-being, concludes independently that he enlarge his herd and graze on more land. By so doing he derives additional income but is only weakly influenced by the effects of overgrazing, at least in the short term. His colleagues do the same. At some point, however, the overgrazing destroys the pasture and disaster strikes all.

Avoiding "Tragedy" is a common challenge that resource managers face around the world, from natural systems like forests and fisheries to human systems like urban space, the telecommunications spectrum, and transportation networks. Harvey Brooks of Harvard University noted that the convenience, privacy, and safety of travel by private automobile encourages each individual to drive to work, school, or stores. At low levels of traffic density, this is a perfectly logical approach to the demands of modern life. At some critical density, however, the road network commons is incapable of dealing with the traffic, and the smallest disruption (a stalled vehicle, a minor accident) dooms drivers to minutes or hours of idleness, the exact opposite of what they had in mind. Examples of frequent collapse of road network commons systems are now legendary, such as traffic congestion in Los Angeles, Tokyo, Naples, Bangkok, and Mexico City.

The common pasture and the common road network are examples of societal systems that are basically local in extent and can be addressed by local societal action if desired. Perturbations to water and air do not follow this pattern, however. The hydrosphere and the atmosphere are examples not of a "local commons" but of a "global commons", a system that can be altered by individuals the world over for their own gain, but if abused can injure all. Much of society's functions are embodied in industrial activity (where the word "industrial" should be interpreted broadly to mean any human action involving the transformation of materials or energy), and it is the relationships among industry, the environment, and society, especially the global commons, that are the topic of this book.

When we objectively view the recent past, one fact becomes clear: the Industrial Revolution as we now know it is not sustainable over time. We cannot keep using materials and resources the way we do now, especially as is the case in the more industrially developed countries.

But what is the alternative? How can we maintain (or improve) our quality of life without degrading or destroying the natural systems on which all life depends?

1.3   Technological Evolution

It is undeniable that modern technology has provided enormous benefits to the world's peoples: a longer life span, increased mobility, decreased manual labor, and widespread literacy, to name a few. Nonetheless, there are growing concerns about the relationships between industrial activity and Earth's environment, nowhere better captured than in the pathbreaking report *Our Common Future*, produced by the World Commission on Environment and Development in 1987. The concerns raised in that report gather credence as we place some of the impacts in perspective. Since 1700, the volume of goods traded internationally has increased some 800 times. In the last 100 years, the world's industrial production has increased more than 100-fold. In the early 1900s, production of synthetic organic chemicals was minimal; today, it is over 225 petagrams (billion metric tons) per year in the US alone! (See the appendix on Units of Measurement in Industrial Ecology for conversion between the metric and imperial number systems.) Since 1900, the rate of global consumption of fossil fuel has increased by a factor of 50, with the effect of the now well-known exponential increase in atmospheric carbon dioxide levels. What is important is not just the numbers themselves, but their magnitude and the relatively short historical time they represent.

Human population growth is, of course, a major factor fueling this explosive industrial growth and expanded use and consumption of materials. Since 1700, the human population has grown ten-fold. While this growth is generally recognized, it is less widely appreciated how closely human population growth patterns are tied to technological and cultural evolution. Indeed, the three great jumps in human population have accompanied the initial development of tool use, the agricultural revolution, and the Industrial Revolution. The Industrial Revolution actually consisted of both a technological revolution, and a "neo-agricultural" revolution (the advent of modern agricultural practices), which created what appeared to be unlimited resources for population growth. Our current population levels, patterns of urbanization, economies, and cultures are now inextricably linked to how we use, process, dispose of, and recover or recycle natural and synthetic materials and energy, and the innumerable products made from them.

Technological evolution generally proceeds in one of two ways. Most of the time technological evolution is incremental, marked by small improvements or changes in existing products or systems that, taken together, improve the quality of life but do not significantly change economic, cultural, or natural systems. In some periods, however, so-called "transformative technologies" change the technological landscape so profoundly that change in the related systems is significant and often difficult. Indeed, economists have identified stages in economic development that can be associated with particular enabling technologies. For example, the introduction of the railroad, the automobile, and electricity changed not just economic and related technological systems, but also culture, national competitiveness, political systems, and most people's way of life at the individual level. It is indeed accurate to say that the railroad was a necessary, and enabling, technology for the rise of Britain as a world economic and political power. It also necessitated other technologies, such as national communication systems using Morse Code, and required accurate timekeeping, thus changing the way time was perceived and measured around the world. Among the other effects of extensive railroad infrastructure was helping to make the American Midwest a viable agricultural enterprise, feeding products to the American East Coast and from there to global markets. That a single technology so transformed vast areas of land and affected the economic wellbeing and personal lives of so many helps make clear the relationship between sustainability and technology.

1.4 The Master Equation

A useful way to focus thinking on the most efficient response that society can make to environmental stresses is to examine the predominant factors involved in generating those stresses. As is obvious, the stresses on many aspects of the Earth system are strongly influenced by the needs of the population that must be provided for, and by the standard of living that population desires. One of the most famous expressions of these driving forces is provided by the "master equation" of industrial ecology:

$$\text{Environmental Impact} = \text{Population} \times [\text{GDP/Person}] \\ \times [(\text{Environmental Impact})/ \\ \{\text{Unit of GDP}\}] \tag{1.1}$$

where GDP is a country's gross domestic product, a measure of industrial and economic activity. This equation has traditionally been called the IPAT equation, where I = impact, P = population, A = affluence (GDP/person), and T = technology (impact/unit of GDP). Let us examine the three terms in this equation and their probable change with time.

Earth's population has risen dramatically over the past several centuries. Thanks to the work of demographers and census-takers the current population is well known, and future populations in each region can be projected based on age structures, birth rates, and death rates. While the rate of population increase has slowed overall, some regions are still growing quickly; the global population is anticipated to peak at nine billion or thereabouts during the 21st century.

The second term in Equation (1.1), the per capita gross domestic product, varies substantially among different countries and regions, responding to the forces of local and global economic conditions, the stage of historical and technological development, governmental factors, weather, and so forth. The general trend, however, is positive, reflecting the aspirations of humans for a better life. Although GDP and quality of life may not be fully connected, we can expect GDP growth to continue, particularly in developing countries.

The third term in the master equation, the degree of environmental impact per unit of gross domestic product, is an expression of the degree to which technology is available to permit development without serious environmental consequences and the degree to which that available technology is deployed.

Although the IPAT equation should be viewed as conceptual rather than mathematically rigorous, it can be used to suggest goals for technology and society. If our aim is to constrain the environmental impact of humanity to its present level (and one could make arguments that we need to do even better than that), we need to look at the probable trends in the three terms of the equation. The first, as discussed above, will likely increase by 20% to 40% before peaking in this half-century. The second term is thought likely to increase over the same time period by a factor of between three and five. Accordingly, to merely hold our environmental impact where it is today, the third term must decrease by something between 50% and 90%. This third term, the amount of environmental impact per unit of output, is primarily a technological term, though societal and economic issues provide strong constraints to how fast this term can change. It is this third term in the equation that is essentially the main focus of the fields of industrial ecology and sustainable engineering and that appears to offer the greatest hope for a transition to sustainable development, especially in the next few decades.

1.5 Facing the Challenge

The 20th century was a period of enormous progress for humanity, progress that was achieved in part by ignoring the possible environmental consequences. The conjunction of inadequately thought-out technological approaches with rapidly rising populations and an increasing culture of consumption is now producing stresses that are increasingly obvious and call out for change: a new perception of our industrialized society.

There are roles for many players in transforming the technology-society-environment relationship. Social scientists need to understand consumption and how it may evolve and be modified. Environmental scientists need to understand how natural systems behave and what are the boundaries of a "safe operating space" imposed by a planet with limited resources and limited assimilative capacity for industrial emissions. Technologists need to develop design and manufacturing approaches that are more environmentally sound. Industrialists need to understand all these frameworks for action and develop ways to integrate the concepts within today's corporate structures. Policymakers need to provide the proper mix of regulations and incentives to promote the long-term health of the planet rather than short-term fixes.

These are great challenges that humanity must face in the 21st century if we are to maintain a healthy and prosperous future for current and future generations. We are finally waking up to the fact that we have pushed many critical natural systems past safe limits, including the climate itself. The field of industrial ecology provides new paradigms, methods, and tools for building a more sustainable civilization, many of which will be presented in this primer. At its core industrial ecology is a vision of a world in which human systems mimic natural systems, working in synergy and without waste.

Further Reading

Chertow, M., The IPAT equation and its variants: Changing views of technology and environmental impacts, *Journal of Industrial Ecology*, *4*(4), 13–29, 2001.

Elhacham, E., L. Ben-Uri, J. Grozovski, Y.M. Bar-On, and R. Milo, Global human-made mass exceeds all living biomass, *Nature*, *588*, 442–444, 2020.

Frischmann, B., A. Marciano, and G.B. Ramello, Tragedy of the commons after 50 years, *Journal of Economic Perspectives*, *23*(4), 221–228, 2019.

Global Energy Assessment (2012). Laxenburg, Austria: International Institute for Applied Systems Analysis.

Hardin, G., The tragedy of the commons, *Science*, *162*, 1243–1248, 1968.

Hooke, R.L., On the history of humans as geomorphic agents, *Geology*, *28*(9), 843–846, 2000.

Janssen, M.A., S. Smith-Heiskers, A.R. Aggarwal, and M.L. Schoon, 'Tragedy of the commons' as conventional wisdom in sustainability education, *Environmental Education Research*, *25*(11), 1587–1604, 2019.

Syvitski, J.P.M., and A. Kettner, Sediment flux and the anthropocene, *Philosophical Transactions of the Royal Society A*, *369*, 957–975, 2011.

Chapter 2

Industrial Ecology Concepts

Chapter Concepts

- Industrial ecology (IE) rejects the concept of waste in favor of a concept of reuse, as is the case in nature.
- Sustainable engineering can be regarded as an operational arm of IE. It provides a template for the environmental and societally sustainable redesign of the modern world.
- Six comprehensive goals and ten operational tools constitute the basic framework of the practice of IE and sustainable engineering.

2.1 From Individual Thinking to Systems Thinking

Since the Industrial Revolution, the activities of organizations large and small have defined much of the interactions between humanity and the environment and significantly shaped our societies. These interactions have, however, traditionally been outside the topics of major significance for corporate decision-makers. Technology's influence on the environment, and especially the potential magnitude of that influence across the full spectrum of economic activity, has generally been under-appreciated in the business world.

Nonetheless, no individual organization exists in a vacuum. Every activity is linked to thousands of processes and transactions and to their environmental and social impacts. A large organization manufacturing high technology products will have tens of thousands of suppliers located all around the world, and the list changes on a daily basis. The organization may manufacture and offer for sale thousands of individual products to a myriad of customers, each of which has her or his own needs and cultural preferences. Each customer, in turn, may treat the product very differently and live in areas with very different environmental characteristics; these are considerations of importance when use and maintenance of the product may be a source of potential environmental impact (e.g., disposing of used motor oil from vehicles). The services and cultural patterns that the product enables will also differ significantly in different communities and societies. When finally discarded, the product may end up in almost any country, in a high technology landfill, an incinerator, beside a road, or in a river that supplies drinking water to local communities.

In such complex circumstances, with social and environmental impacts at many scales, how has industry approached its relationships with the outside world? Satisfying the needs of its customers has always been common practice, at least in market economies. Industry has, however, been less adept at identifying some of the consequences, especially the long-term environmental consequences, of the ways in which it goes about satisfying needs. Since the beginning of the 1970s, when modern environmentalism began to arise, analysis of human-environment interactions has improved to the point where we can now see how past technological "solutions" have caused widespread damage. Examples of a few of these interactions have been collected by Dr. James Wei of Princeton University (adapted and displayed as Table 2.1), indicating the peril of

Table 2.1.   Relating current environmental problems to yesterday's industrial responses.

Yesterday's need	Yesterday's solution	Today's problem
Non-toxic, non-flammable refrigerants	Chlorofluorocarbons	Ozone hole
Automobile engine knock	Tetraethyl lead	Widespread lead poisoning and contamination
Locusts, mosquitos, malaria	DDT	Adverse effects on birds and mammals
Expanded food production	Synthetic fertilizer	Eutrophication
Food safety and preservation	Plastic packaging	Macroplastic and microplastic contamination

The old idea: There are "natural"
areas and there are "industrial" areas

The new idea: technology sits
within the natural system

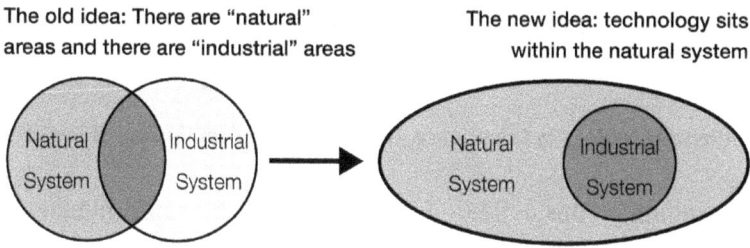

Figure 2.1. The transformation from natural and industrial systems as essentially independent entities to the realization that technology is embedded within the natural system.

unintended consequences in a world in which industrial operations were perceived as essentially unrelated to the wider world, as suggested in the left side of Figure 2.1.

It is important to note that the relationships in Table 2.1 were not the planned result of disdain for the external world by industry. Several of the solutions were, in fact, great improvements over the practices they replaced, and their eventual consequences could not have been forecast with any precision. What was missing, however, was any structured attempt to relate the techniques for satisfying customer needs to possible environmental consequences.

2.2 Defining Industrial Ecology

The broad approach to understanding and managing industry-environment-sustainability interactions is termed "industrial ecology" (IE). The overall concept is, in part, technological. As applied in manufacturing, IE involves the design of industrial processes, products, and services from the perspectives of product competitiveness, environmental concerns, and society. IE is also, in part, sociological. In that regard, it recognizes that human culture, individual choice, and societal institutions play major roles in defining the interactions between our technological society and the environment. It recognizes as well that modern technological and societal systems are fully connected with and embedded within the natural world, as indicated at the right side of Figure 2.1.

IE focuses in many different ways on topics that ultimately are connected to and enabled by physical resources. Because of this focus,

Gjalt Huppes of Leiden University proposes a succinct definition of the field:

Industrial ecology is the science of the physical functioning of society.

Expanding on this definition, IE is the means by which humanity can deliberately approach and maintain sustainability, given continued economic, cultural, and technological evolution. The concept requires that an industrial system be viewed not in isolation from its surrounding systems, but in concert with them. It is a systems view in which one seeks to optimize the total materials cycle from virgin material, to finished material, to component, to product, to obsolete product, and to ultimate disposal, and to carefully evaluate the impacts of the industrial system on the environment.

In this definition, the emphasis on *deliberate* differentiates the IE path from unplanned, precipitous, and perhaps quite costly and disruptive alternatives. By the same token, the definition indicates that IE practices have, or should have, the potential to support a sustainable world with a satisfactory quality of life for all.

Practitioners of IE interpret the word *industry* very broadly: It is intended to represent the sum total of human activity, encompassing mining, manufacturing, agriculture, construction, energy generation and utilization, transportation, product use by customers and service providers, infrastructure systems, service networks, and waste disposal. IE is not limited to the domain within the factory walls but extends to all the impacts on the planet resulting from the presence and actions of human beings, and thus encompasses society's use of resources of all kinds.

Key concepts of this approach include conservation of mass (all material must be accounted for), conservation of energy (all energy must be accounted for), and consideration of the technological arrow of time — the realization that as society becomes more technologically advanced it builds on its past technological base and so cannot sustain or improve itself without strong reliance on technology.

One of the most important features of IE is that it rejects the concept of waste. Dictionaries define waste as useless or worthless material. In nature, however, nothing is eternally discarded — in various ways almost all materials are reused, often with great efficiency. Natural systems have evolved these patterns because acquiring materials from their reservoirs is costly in terms of energy and resources, and thus something to be avoided whenever possible. In our industrial world, discarding materials wrested from the Earth system at great cost is also generally unwise. Hence, materials and products that are obsolete should be termed *residues* rather than *wastes*. It should be recognized that wastes are merely residues that our economy has not yet learned to use efficiently.

A full consideration of IE includes the entire scope of economic activity: mining, agriculture, fishing, forestry, manufacturing, transportation, all manner of services, waste management, and the consumer behavior that drives demand, and so forth. It is obviously impossible to cover the full scope of IE in one book, but in this short volume we present the basic concepts and tools of IE, presented alongside examples of seminal work in the field.

2.3 Toward Sustainable Engineering

Engineering has traditionally been regarded as the specialty that employs scientific principles to achieve practical ends. Because engineers are problem-solvers in the context of their cultures, resources and disposal sites were historically (and conveniently) regarded as limitless. As a result, engineering designs made profligate use of resources and unintentionally caused a great deal of environmental damage. Moreover, considerations of the social implications of corporate professional actions tended to be limited to issues that directly affected the use of products or designs. These approaches are now clearly recognized as outdated, and modern engineers acknowledge the need to do better. But how?

It is vital to recognize that our current circumstances require the transformation of the engineering profession toward the practice of *sustainable engineering*. This new paradigm seeks to deliver solutions that consider not just performance and cost, but also social and community impacts, resource use, and environmental implications. Engineering projects tend to have a narrow focus on a location or unit, but sustainable engineering also asks, "How far do we need to go to improve the overall *system*?" This latter question is the province of industrial ecologists, who study the interactions between technology and the wider world. Thus, sustainable engineering can legitimately be regarded as an operational arm of IE, and the essence of IE and sustainable engineering can be briefly stated:

Industrial ecology and sustainable engineering provide a template for the comprehensive redesign of the modern world with the goals of better living for all and decreased impacts on the natural world.

A template is particularly useful only if those who use it understand and address the characteristics of the system they are attempting to redesign. It is useful, therefore, to present some of the characteristics of modern technology:

- *Technology is uncertain* (the best solution is never obvious, and experimentation is vital);
- *Technology is progressive* (change occurs by evolution and transformation);
- *Technology is analytical* (it measures actions and new ideas);
- *Technology is cumulative* (it builds on previous knowledge and existing capabilities);
- *Technology is systemic* (interdependence of technologies is required for progress);
- *Technology is embedded* (technology sits within natural systems);
- *Technology is accelerating* (the waves of technological transitions are ever shorter).

Any attempt to refashion technology must be responsive to these characteristics and adapt as they change, else the attempt will be doomed to failure. IE provides tools for understanding and quantifying how changes to technology will affect the environment at a systems level.

In 2002 William McDonough and Michael Braungart published a small book titled "Cradle to Cradle" in which they proposed that a sensible goal for society (and thus for sustainable engineering) was not to be *less bad* but rather to be *good*, advocating a philosophy centered on reductions in material use, a focus on reuse, and a commitment to recycling. These are important aspirational goals but say little about how those goals might be achieved. That part of the task belongs to IE and sustainable engineering.

Table 2.2. Aspects of industry–environment interactions.

Activity	Time	Focus	Endpoint	Corporate view
Remediation	Past	Local site	Reduce human risk	Overhead
Treatment, disposal	Present	Local site	Reduce human risk	Overhead
Industrial ecology	Future	Global	Sustainability	Strategic

Industrial environmental management has come a long way over the past 50 years. The contrast between traditional environmental approaches to industrial activity and those suggested by IE can be demonstrated by considering several time scales and types of activity, as shown in Table 2.2. The first topic, remediation, deals with such things as removing toxic chemicals from soil. Remediation concerns past mistakes, is very costly, and adds nothing to the productivity of industry. The second topic, treatment, storage, and disposal, deals with the proper handling of residual streams from today's industrial operations. The costs are embedded in the price of doing business, but accomplishing this task contributes little or nothing to corporate success except to prevent criminal actions and lawsuits. In contrast, IE and sustainable engineering deals with practices that look to the future and seeks to guide industry to cost-effective and planet-friendly methods of operation. Doing so will render more nearly benign an industry's interactions with the environment and will optimize the entire manufacturing process for the general good (and, very often, for the financial good of the corporation).

A meritorious example of sustainable engineering is that of Interface, Inc. of Lagrange, Georgia, US, a maker of floor coverings. A recent carpet designed by that corporation and marketed for use in offices, airports, and other high-traffic areas does not only use less energy during its production than the carpets of its competitors, it is actually carbon negative (that is, its materials store more carbon than is consumed in manufacture). The production process and its philosophy are described by Gertner (see Supplementary Reading). This approach is technology at its best — helping to sustain the planet rather than damaging it to a lesser degree than its competitors.

2.4 The Goals and Tools of Industrial Ecology

As in any field, there are key questions in IE. Unlike biological ecology, we are interested not only in the functioning of the technological system *per se*, but also in the industrial ecosystem's interactions with and implications for the natural and social systems of the planet. We specifically concentrate on a single species (humans), our relationship with the environment, and the impacts of technological operations and choices on social systems. From this broad framework, here is a set of key questions that this book explores:

(1) How do modern technological cycles operate?
  (1.1) How are industrial sectors linked?
  (1.2) What are the environmental and social opportunities and threats related to specific technologies or products?
  (1.3) How are technological products and processes designed, and how might those approaches be usefully modified?
  (1.4) Can ideal material cycles, from extraction to final disposal, be established for the technological materials used by our modern society?
  (1.5) How do technological cycles interact with culture and society, and what are the implications inherent in these "second order" effects of technology?
(2) How do the resource-related aspects of cultural systems operate?
  (2.1) How do corporations manage their interactions with the environment and society, and how might corporate environmental management evolve?
  (2.2) How can the culture/consumption influence on materials cycles be modulated?
  (2.3) How can engineers appreciate their relationships with environment and society?
  (2.4) How might IE systems be better understood?
(3) What are the limits to the interactions of technology with the world within which it operates?
  (3.1) What limits are imposed by non-renewable, non-fossil resource availability?
  (3.2) What limits are imposed by the availability of "green" energy?

(3.3) What limits are imposed by the availability of water?

(3.4) What limits are imposed by environmental and/or sustainability concerns?

(3.5) What limits are imposed by institutional, social, and cultural systems?

(4) What is the future of the technology–environment–society relationship?

(4.1) What scenarios for development over the next several decades form plausible pictures of the future of technology and its relationship to the environment and social systems?

(4.2) Should systems degraded by technological activity be restored, and if so, how?

These are broad questions that get to the nature of the basic physical relationships between humans and our environment. A complementary approach is to consider what overall goals IE should address. Although opinions could well differ regarding this list of goals, here are four that will be addressed in this book:

Goal 1: Characterize and quantify material flows and stocks in order to understand our changing demand for resources, how they might best be supplied, and the potential for circularity.

Goal 2: Characterize and quantify the environmental impacts of modern technologies and their uses, in order to redesign them for lower impacts and positive outcomes.

Goal 3: Describe how physical flows move through the economy and among organizations, in order to enable effective supply chain management and utilization of resources.

Goal 4: Understand the evolving relationship between humans and technology and the role of human behavior and psychology in effective management of technological systems.

You will learn how to address these goals and work toward a more sustainable future in the following chapters.

Further Reading

Allen, D.T. and D.R. Shonnard, *Green Engineering: Environmentally Conscious Design of Chemical Processes*, Upper Saddle River, NJ: Prentice Hall, 2002.

Ayres, R.U. and L.W. Ayres, *A Handbook of Industrial Ecology*. Cheltenham, UK: Edward Elgar, 2001.

Frosch, R.A., Industrial ecology: A philosophical introduction, *Proceedings of the National Academy of Sciences of the U.S.*, *89*, 100–103, 1992.

Gertner, J., Better living through CO_2, *New York Times Magazine*, pp. 27–33 and 61, June 27, 2021.

Graedel, T.E. and J.A. Howard-Grenville, *Greening the Industrial Facility*. New York: Springer, 2005.

McDonough, W. and M. Braungart, *Cradle to Cradle: Remaking the Way We Make Things*. North Point Press, 2010.

Chapter 3

The Linked Systems of Human Society and Industrial Ecology

Chapter Concepts

- The material footprints of modern societies accelerated rapidly in the mid-20th century, and show no signs of leveling off.
- Consumption is ultimately the product of individual decisions, and sustainable levels of consumption at the societal level can only be achieved by influencing individual choice.
- Attempts to measure social progress in different countries suggest that material stocks and energy use are not directly related to perceived quality of life, thereby suggesting that a less material-oriented world might become one that is accepted and valued by most people worldwide.

3.1 Historical Linkages of Humans and Resources

Throughout early human existence, a hunter/gatherer lifestyle was the workable approach to life and society. Somewhat later, agrarian lifestyles became the norm. Those who live in countries that are technologically advanced tend not to realize that a hunter/gatherer or agrarian lifestyle remains the case for much of Earth's people, and remote regions still retain societal groups who continue to live as hunter/gatherers. It is anticipated, however, that the next few decades will witness an unprecedented

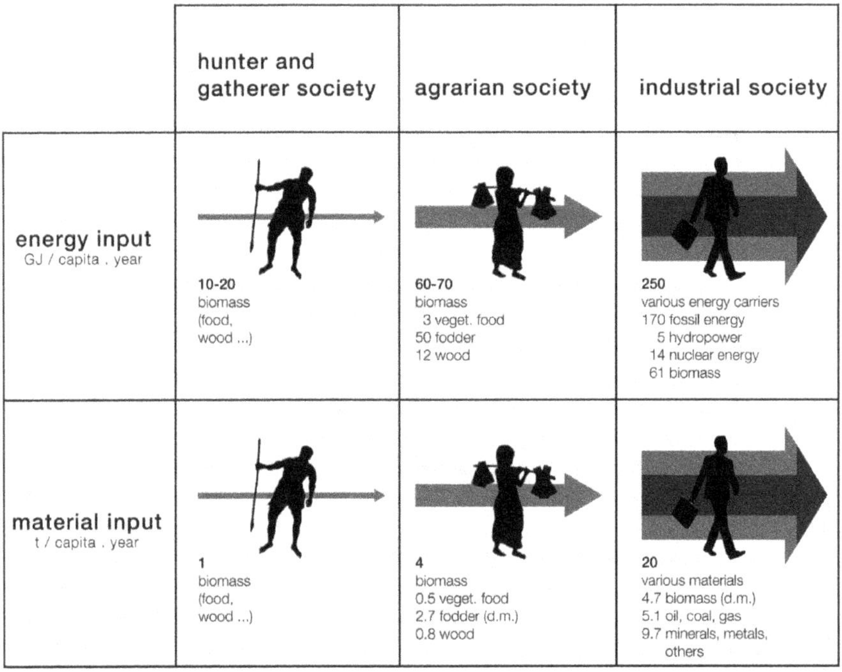

	hunter and gatherer society	agrarian society	industrial society

energy input
GJ / capita . year
	10-20
	biomass
	(food,
	wood ...)
		60-70
		biomass
		  3 veget. food
		  50 fodder
		  12 wood
			250
			various energy carriers
			  170 fossil energy
			  5 hydropower
			  14 nuclear energy
			  61 biomass

material input
t / capita . year
	1
	biomass
	(food,
	wood ...)
		4
		biomass
		  0.5 veget. food
		  2.7 fodder (d.m.)
		  0.8 wood
			20
			various materials
			  4.7 biomass (d.m.)
			  5.1 oil, coal, gas
			  9.7 minerals, metals,
			  others

Figure 3.1. Typical resource consumption by the three traditional sociometabolic regimes (adapted from Fischer-Kowalski and Haberl, 1998).

level of hunter/gatherer to agrarian to industrial transition, especially as more and more people move from rural to urban areas. What might be the implications of this process?

Some clues are given in Figure 3.1, in which the energy and material properties of hunter/gatherer, agrarian, and industrial lifestyles are contrasted. It is striking that the material use of industrial societies is some five times that of agrarians and 20 times that of hunter/gatherers. The main sources of energy also underwent a marked transition over time — from an almost total dependence on biomass in the agrarian regions to a strong emphasis on fossil fuels during industrialization to the advent of renewable and transportable energy on demand in the near future. Many opportunities have arisen as a consequence of this evolution. One is that fossil fuel energy permitted mechanized farming, which freed much of the labor force for other employment. Another is that the increased population density enables public transportation, better schooling, and improved health care. Liabilities come along with the transition as well, however: a changed social structure and the spatial concentration of pollution sources, to name only two.

Human appropriation of resources is not only a function of technology but also social and cultural drivers of consumption. The flows of materials, energy, and emissions in a society is called its *socio-economic metabolism* (SEM). The higher the metabolic rate the more resources per inhabitant are needed, the more outflows of wastes and emissions are produced, and the higher is the environmental impact. SEM enlarges traditional biological definitions of metabolism to analyze human communities and industry, often at a national level. Thus, just as the metabolic analysis of a bird can be expanded to include the nest it builds, so can that of society address aqueducts and railroads.

3.2 Accelerating into the Anthropocene

Modern humans use more resources per capita than was the case historically and Earth's human population is increasing, so it would be reasonable to expect that global resource use would be increasing as well. And, because many features of resource use are related to population and income, one would expect such specific metrics as fertilizer consumption and energy use to be increasing. Additionally, if environmental impacts are a consequence of such actions, those impacts probably have also increased. But by how much?

This question was addressed by a group of researchers under the auspices of the International Geosphere-Biosphere Program (IGBP). The results for a dozen globally-aggregated socio-economic indicators are shown in Figure 3.2. The data, extending where possible back to 1750 at the beginning of the Industrial Revolution, communicate their message in a quick glance: Not only are all the indicators increasing over time, but starting in about 1950 they began to rise quite rapidly. The sharp mid-20th century transition is mirrored in the trends for a variety of Earth system indicators.

Socio-economic trends

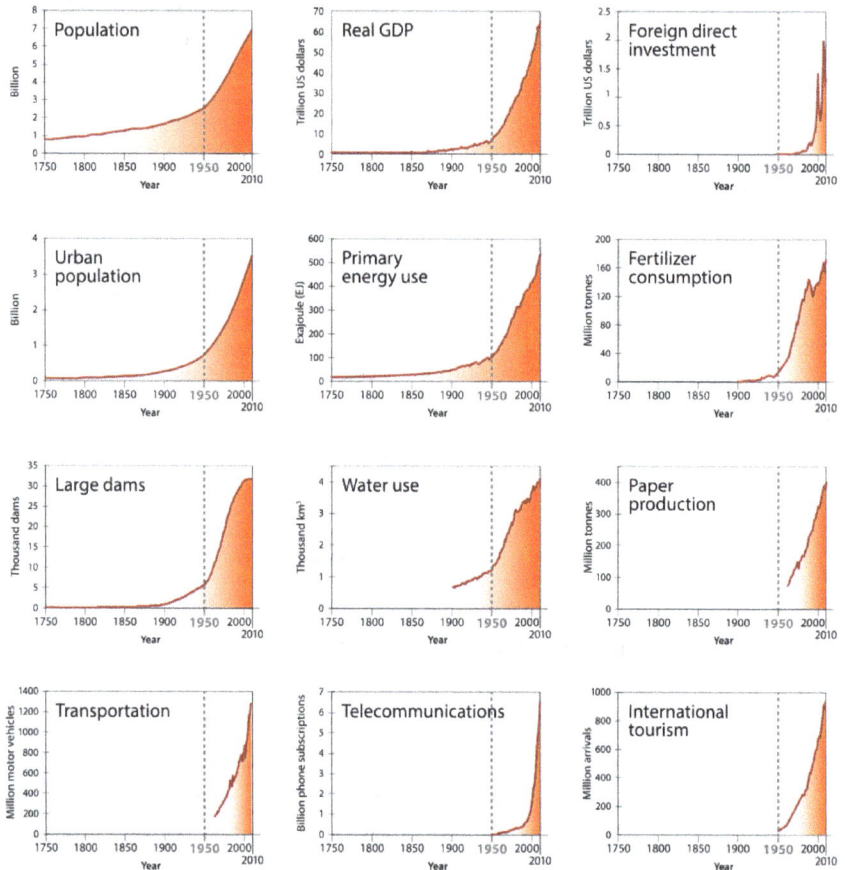

Figure 3.2. (a) Socio-economic and (b) earth system trends, 1750–2010 (Steffen *et al.*, 2015).

The impact of this work on socio-economic and Earth System trends was so dramatic that it inspired a new label for the post-1950 period: *The Great Acceleration*. The graphs of Figure 3.2 visually demonstrate what a shorter-term material flow analysis or environmental assessment cannot — that the

Earth system trends

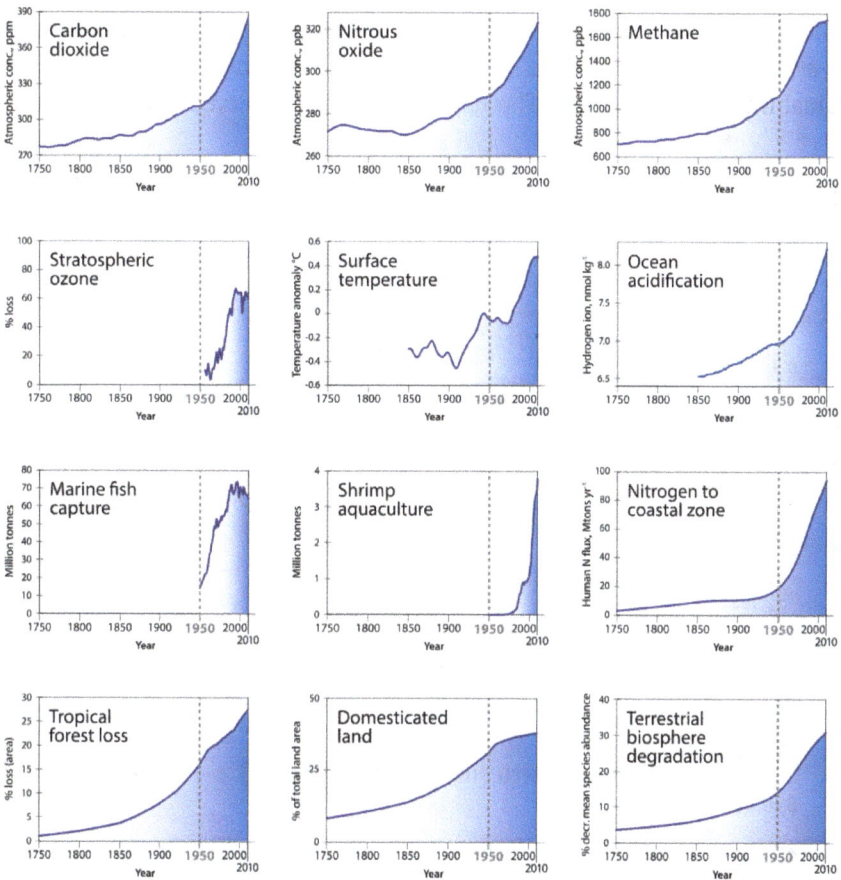

(b)

Figure 3.2. (*Continued*)

impact of humans on the planet has become so amazingly large as to instigate panicky thoughts about the prospects for our planet and society. Indeed, the mark of humans on planetary systems has become so dominant that we are clearly living in the Anthropocene.

3.3 Evolutionary Approaches to Human Wants and Needs

The picture painted above regarding human desires and their satisfaction is a highly problematic one so far as the environment and long-term sustainability are concerned. The natural world has evolved over time to be

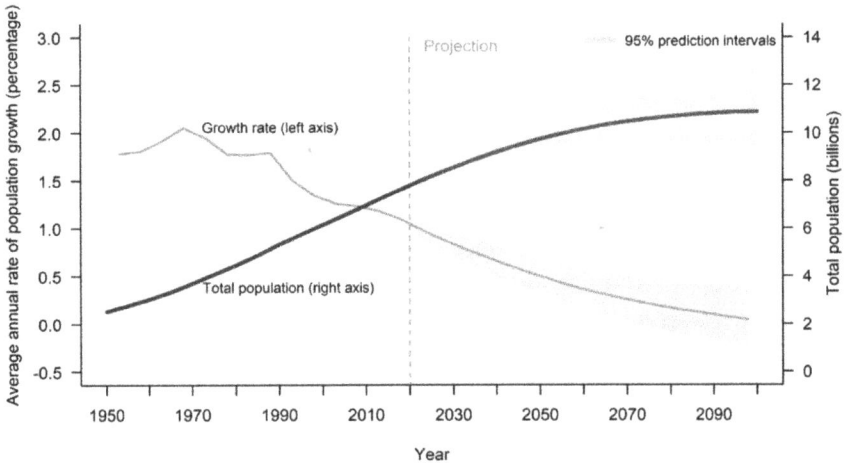

Figure 3.3.   Global population trends (United Nations Population Division).

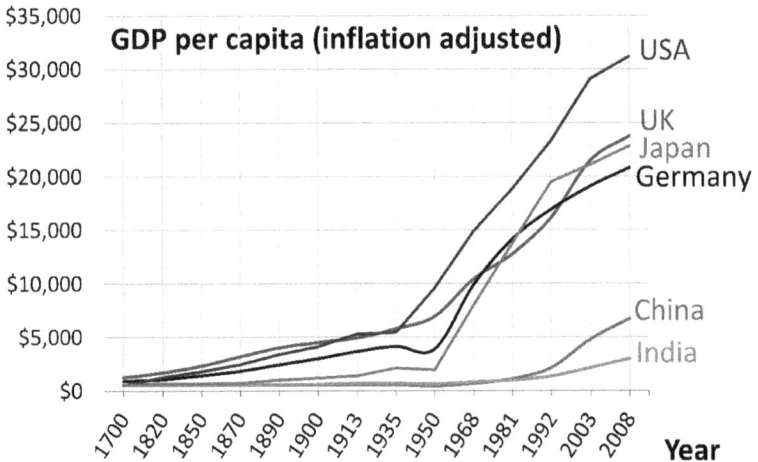

Figure 3.4.   The evolution of gross domestic product per capita for the period 1700–2008 (Wikipedia).

cyclic in resource use, or nearly so, with populations of species constrained by available habitats and resources. The analogous human picture is divergent instead of cyclic, as can be seen by considering the IPAT equation of Chapter 1. Figure 3.3 shows that the human population is predicted to grow for the rest of this century, stabilizing by perhaps 2100: an encouraging if long-term sign. However, Figure 3.4 exemplifies that gross domestic product per capita, a reliable measure of affluence, continues to increase (with moderate economic fluctuations) all over the world.

If **I** (the environmental impact term of the IPAT equation) is likely to be stressed by increasing affluence (**A**) in the coming decades, and by population (**P**) expansion for many years to come, there is a big job ahead for technology (**T**): to serve the needs and desires of billions of people while sharply decreasing the related environmental implications. Transitioning from technology's use of fossil fuels is surely part of what is needed, but much more dramatic initiatives will be required to respond to the challenge, and those initiatives will need to be sociological as well as technological because the former drives the latter.

Further Reading

Fischer-Kowalski, M., and H. Haberl, Sustainable development: Socioeconomic metabolism and the colonization of nature, *International Social Science Journal, 158*, 573–587, 1998.

Steffen, W., W. Broadgate, L. Deutsch, O. Gaffney, and C. Ludwig, The trajectory of the Anthropocene: The great acceleration. *The Anthropocene Review, 2*, 81–98, 2015.

UN Statistics Division, Population Facts, 2019.

Chapter 4

Living Within Limits of Natural Systems

Chapter Concepts

- For millennia humans have been employing nature's bounty with little or no appreciation of the fact that the gifts provided by nature have limits.
- Industrial ecologists measure the pressures on nature by three approaches: sustainable yield assessments, carrying capacity evaluations, and planetary boundary exceedances.
- The United Nations measures human progress by the Sustainable Development Goals and the degree to which those goals are being met. Many of the goals are closely related to industrial ecology concerns and methodologies.

4.1 Human and Natural Systems are Intertwined

Imagine for yourself the most remote location on the planet, one so far from population centers or shipping routes that it seems free of human impact. Now go to that location, set up your analytical instruments, and take samples of the air and water. You will readily find molecules synthesized only by modern industry, never by nature: the footprints of humanity. Or, go to a location as intensely human as you can imagine, one showing no indications (other than human beings) of "nature": a skyscraper in a major city, an automobile assembly line, a museum of modern art.

Observe with an ecologist's eye, or take samples of a few surfaces and examine them with a microscope. You will readily find plants, bacteria, insects, and perhaps small mammals and birds: the footprints of nature. Human and natural systems are intertwined at a global scale; we cannot understand the behavior of one without accounting for the influence of the other.

It may seem obvious to those of you reading this book that the actions of humans affect the planet and vice versa, but the ethos of human development has often been to treat them separately, especially since the Industrial Revolution. Nature was often seen in the past as a limitless store of natural resources that exists for our benefit, to do with as we wish. Throughout history we have built vast technological systems to harness nature's services, to provide food, water, energy, and materials, and to absorb our wastes, without much consideration of the consequences of how human actions affect natural systems, or how they affect ourselves. We have simply assumed that the rivers will keep flowing, plants will keep growing, and that if we run out of something locally, we can go find what we need somewhere else. In this mindset, natural systems can be tamed completely through human ingenuity and engineering prowess.

But obviously the planet is not limitless in its land, resources, or ability to cope with exploitation. Modern environmentalism was sparked by a series of major events when natural systems became so degraded that they essentially shut down and their effects on humans became all too clear. For example, the proliferation of gasoline-powered tractors and inappropriate farming practices in the American plains coupled with severe drought led to the Dust Bowl in the 1930s. In some regions 75% of all the topsoil was lost, ruining agriculture and causing the largest internal migration in American history. In Cleveland, Ohio, a major manufacturing hub, the local Cuyahoga river became so laden with industrial pollution that it famously caught fire more than a dozen times. And across the world, the effects of air pollution on human health were sadly demonstrated during acute events such as the London "Great Smog" in the 1950s that killed thousands of people. These events (and many others in history) clearly demonstrated that natural systems are finite and can fail.

Natural systems are typically complex. How they will respond to human perturbations is not always predictable and is often nonlinear.

A wetland ecosystem may be able to naturally filter pollutants and maintain good water quality up to a point, but may be overwhelmed and stop functioning once the pollutant loading becomes too great. However, complex systems also have the ability to adapt to changing circumstances. This "adaptive capacity" enables many natural systems to heal themselves, given enough time. For hikers enjoying a walk in the seemingly pristine woods in the Northeast USA, it is hard to fathom that nearly all of that land was clear-cut in the 1800s. The forests that have regrown there are very different in composition compared to what was there before, but they are still forests, providing habitat for wildlife, supplying oxygen and clean water, and providing quiet space for reflection.

Humans are a part of nature and, like any organism, need to utilize resources from the environment to survive. But how are we to know when we are using too many resources, when the natural systems we rely upon are in danger of losing their adaptive capacity, and when our actions pose a danger to ourselves?

4.2 Sustainable Yield, Carrying Capacity, and Planetary Boundaries

One of the central challenges posed by sustainability is how to "live within our means" in a planetary sense, to ensure that we can "[meet] the needs of the present without compromising the ability of future generations to meet their own needs", in the words of the seminal 1987 report *Our Common Future*, also known as the Brundtland report. These words convey imperative general goals but must be operationalized and quantified in order to be useful.

Sustainable yield is an important concept and metric that is used for natural systems that provide us with physical resources, such as aquifers that bring water or forests that can provide timber. Aquifers can be replenished by groundwater infiltration, and trees can obviously grow back in forests following cutting, so some uses of these resources will not cause these natural systems to fail. Sustainable yield is defined as the amount of resources that can be taken while still maintaining the normal functioning of an ecosystem. If the sustainable yield is exceeded, then the natural system will need time to recover, during which harvesting or extraction may need to be halted or greatly diminished.

Determining a sustainable yield is necessary to solving the Tragedy of the Commons dilemma presented in Chapter 1. As long as the extraction of resources can be controlled, thus allowing limited extraction to occur up to the maximum sustainable yields, both current and future generations can benefit from the continued healthy functioning of natural systems. Many fisheries around the world that had been in dire conditions due to collective overfishing have recovered as a result of instituting quotas built around their sustainable yields.

Clearly oceans do more than supply fish, and forests do more than supply trees. Humans and all organisms need to utilize a broad range of resources from their environments. An ecosystem can only support so many individuals before the extraction of resources exceeds the sustainable yield. This maximum population is called the 'carrying capacity' of the system. Carrying capacity can be estimated for small, local ecosystems all the way up to planetary scales. Scientists have long worked to estimate the human carrying capacity of Earth. The eminent Harvard biologist E.O. Wilson puts the number at 9–10 billion humans, but estimates range widely, in part because the sustainable yield of so many systems is unknown, as are the dynamics among them.

Just because we are currently below Wilson's estimated carrying capacity for Earth does not mean that everything is okay. Some systems are highly degraded and seem headed for catastrophe, while others are less disturbed. A final concept to introduce here is called "planetary boundaries". Similar to the idea of sustainable yield, these are thresholds past which natural systems on Earth can no longer maintain themselves and become degraded, sometimes irreversibly. While yields focus on extraction of resources, planetary boundaries can also consider Earth's capacity for absorbing pollution and other impacts from human activities.

The concept of planetary boundaries was introduced in a seminal article in 2009 by a large group of Earth and environmental scientists headed by Johan Rockström. In it they defined nine planetary systems and estimated the planetary boundary for each, as well as whether or not the boundary was being exceeded at that time. Figure 4.1 reproduces the main figure from the article. The estimates in this research were semi-quantitative but

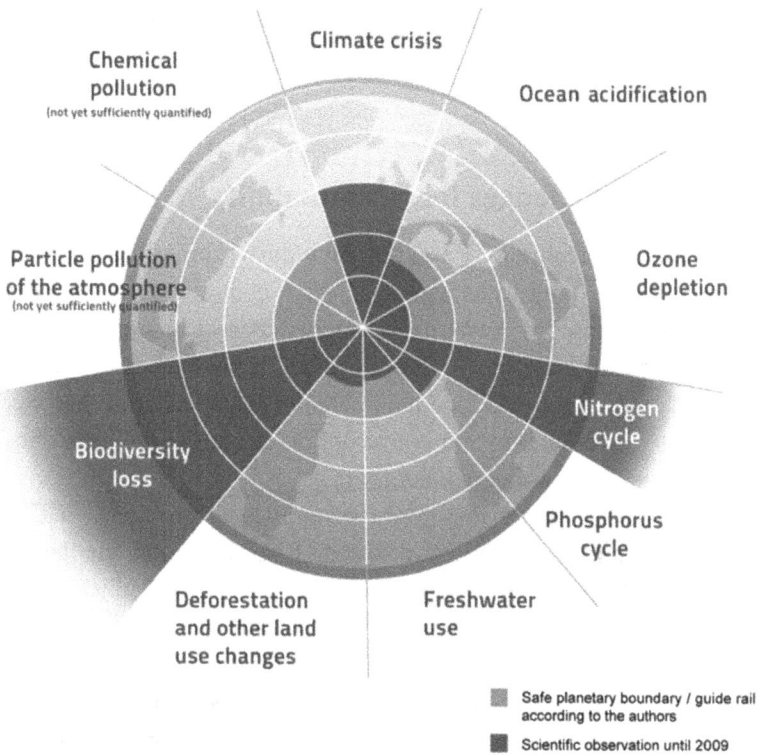

Figure 4.1.   Planetary boundaries (Rockström *et al.*, 2009).

show clearly that, in their judgment, human activities are causing much more harm to some systems than others. Climate crisis is one of the categories that is exceeding planetary boundaries, with dire consequences predicted if we fail to mitigate emissions or adapt to the effects of climate change. But the climate crisis is not the only crisis.

In a 2023 update, Rockström and colleagues reassessed the planetary boundary designations and status of the boundaries, finding that nearly all are already exceeded (see Further Reading). These evaluations are challenging and the results contentious, but there is little doubt that they reveal a planet under extreme stress.

The concepts of sustainable yield, carrying capacity, and planetary boundaries help set the stage for the use of many industrial ecology methods and analyses. Being able to quantify these thresholds is central to assessment tools such as criticality assessment, environmental foot-printing, or life cycle impact assessment, as we will see in future chapters.

4.3 Grand Challenges for Sustainability

Advancements in technology and human industry, broadly speaking, have been hugely beneficial to us as a species. A central theme has been that advances in technology have allowed us to extract resources more efficiently, spreading material wealth but at potentially great cost to natural systems. It is clear that we face many challenges ahead, and that our current models of production and consumption are not sustainable. So, how should we focus our collective attention?

Grand challenges are long-term, systems-level problems that require global action and cooperation to solve. Lists of grand challenges are released by different institutions periodically. In 2019, the National Academy of Engineering (NAE) in the United States released a report titled *Environmental Engineering for the 21st Century: Addressing Grand Challenges*, which laid out what the science and technology establishment viewed as the areas needing the most attention and having the most promise to improve the environmental sustainability of our civilization. They were:

- Sustainably supply food, water, and energy;
- Curb climate change and adapt to its impacts;
- Design a future without pollution and waste;
- Create efficient, healthy, resilient cities;
- Foster informed decisions and actions.

Solving each of these challenges necessitates understanding the complex relationships between human and natural systems and using this understanding to design our farms, cities, and all of human industry to be as resource efficient, safe, and equitable as possible. This is the vision of industrial ecology.

In this book, you will learn about industrial ecology principles, methods, and tools that are used to tackle these grand challenges. Life cycle assessment and network modeling reveal how food–energy–water systems are interdependent and provide metrics for tracking whether production is in fact 'sustainable'. Input–output analysis is used to track how carbon emissions occur throughout the global economy, while scenario modeling allows us to consider how technology, development, and population trends may affect future emissions. Closing materials loops is a central principle of industrial ecology. Material flow analysis provides a quantitative picture of our physical economy, while design strategies and new circular economy business models lay out frameworks for delivering innovative solutions. There are many inspiring case studies from around the world that demonstrate successful resource cycling strategies and can serve as models for the future, from islands to industrial parks to innovative urban designs.

Grand challenges are by their nature complex and involve many trade-offs and unintended consequences. You will learn throughout this book that the same topic can be analyzed in multiple ways. The oft-used metaphor is that industrial ecology provides different tools in your toolbox, tools that can be used to generate results and provide interpretation to designers and policymakers who are responsible for making decisions and taking action. Sustainability is inherently multi-disciplinary and multi-factorial; it is not just about carbon emissions, or toxicity, or equity, or efficiency. So, sustainability frameworks like industrial ecology need robust methods for interpreting multiple metrics and types of results. As the NAE list suggests, developing effective communication and decision methods is itself a grand challenge! Across the tools and topics covered elsewhere in this book you will see how industrial ecologists have grappled with the complexity and multi-dimensionality of sustainability through both quantitative and visual means.

Further Reading

Brundtland, G.H., Chairperson, *World Commission on Environment and Development, Our Common Future.* Oxford: Oxford University Press, 1987, ISBN 019282080X.

National Academies of Sciences, Engineering, and Medicine. *Environmental Engineering for the 21st century: Addressing Grand Challenges.* Washington, DC: National Academies Press, 2019.

Roser, M., Historical index of human development (Generated in 2018), https://ourworldindata.org/human-development-index, accessed August 25, 2021.

Rockström, J. *et al.*, A safe operating space for humanity, *Nature*, *461*, 472–475, 2009.

Rockström, J. *et al.*, Safe and just Earth system boundaries, *Nature*, *619*, 102–111, 2023.

United Nations, Sustainable Development Goals, 2015, https://sdgs.un.org/goals.

Wilson, E.O., *The Future of Life*. Alfred A. Knopf, 2002.

The Methodology of Industrial Ecology

Chapter 5

Providing Services to Society:
In-Use Stocks

Chapter Concepts

- Physical stocks of materials in durable products and infrastructure are used to provide long-term services to society.
- In-use stocks are predominantly functions of population, income, and level of urbanization.
- Once materials are in use as stocks, they often remain in use for decades or generations.

5.1 Standing Stocks

The first measure of ecosystem function for an ecologist is often the *standing stock*: the amount (by number, mass, or some other convenient metric) of a particular species or species group in a particular area. This is generally done by sample collection and species counting but can also involve semi-quantitative estimates based on sources of various kinds. Figure 5.1(a) illustrates the results of research on global biomass. Figure 5.1(b) provides increased detail for groupings of animal species. The results demonstrate that plants (mostly living on land) dominate global biomass, that animals are a small fraction of the total, and that humans are a small portion of the animal fraction. Such results provide the

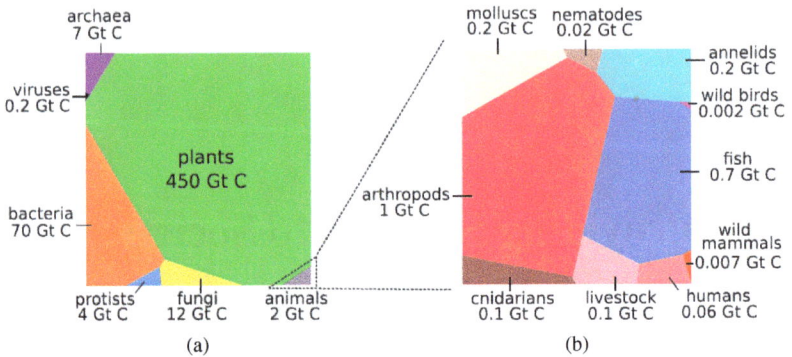

Figure 5.1.   (a) Absolute biomass of the global distribution of biomass by kingdom. (b) Absolute global biomass of different animal taxa (Bar-On *et al.*, 2018).

basis for more extensive studies, such as the rate of change of different biomass groupings over time.

In an analogous fashion, similar approaches can be used to quantify the standing stock of material that has been mobilized and fashioned by humans — this is generally termed the *in-use stock*. In this chapter, we illustrate several results from this branch of industrial ecology and discuss the information that an in-use stock analysis provides.

5.2 The Value of Material Stocks

In previous chapters, we have suggested the ways in which extraction, processing, manufacture, use, and discard create multiple interlocking flows of materials. We have also mentioned that those flows, operating at different time scales, generate in-use stocks of materials. In this chapter, we take a more quantitative look at those stocks, the stimuli that ultimately create the stocks, and the environmental implications of stock existence and use.

Humans do not mine and process materials because mining and processing are their ultimate purposes. Rather, they become miners and processors because humans value the benefits derived from materials when they become stocks in use. This reliance on stocks is easy to demonstrate. In Figure 5.2, we reproduce the results of a study that plots per-capita

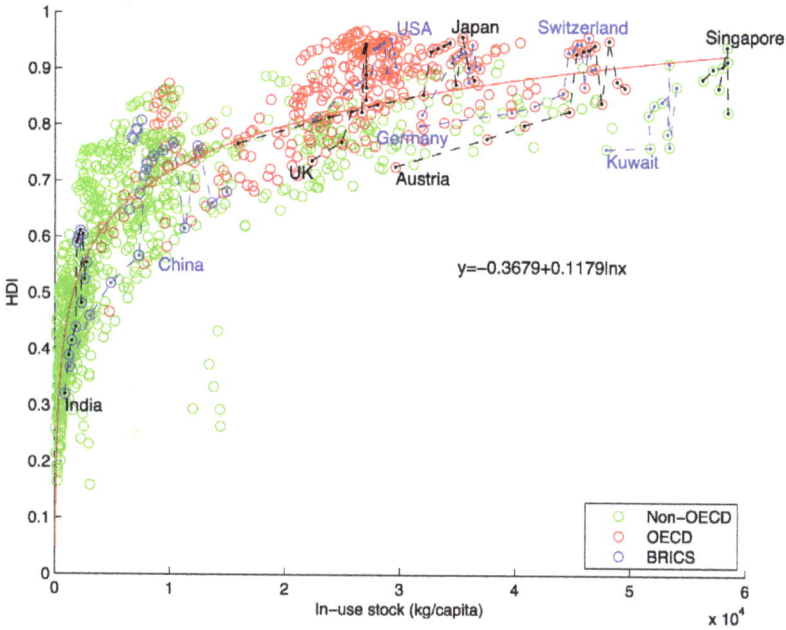

Figure 5.2.   Human Development Index (HDI) as a function of per capita in-use stock (Lin *et al.*, 2017).

in-use stocks (actually the in-use stocks of iron, aluminum, and concrete, generally the three largest mass flows) in most of the world's countries vs. the human development index (HDI), an index generated by the United Nations Development Program based on three components: (1) health (measured by life expectancy at birth), (2) education level (determined by schooling achieved and expected), and (3) living standards (measured by income per capita). It is immediately clear that those with higher HDI rank possess a much greater amount of in-use material stocks than those whose HDIs are lower; that is, as people become wealthier and better educated, they acquire more stuff. For those people, housing is larger and the associated amenities are more numerous, private transport is common, and public services such as roads, bridges, electronic systems, and the like are routinely available.

It is also clear from Figure 5.2 that those with HDI levels above about 0.8 can have a very wide range of in-use stock — the level for Singapore is around twice that of the USA and Germany, for example. This may be due to geographical differences (Singapore is a densely populated island, for example) or spatial patterns (which influence the per capita level of public transportation). In any case, it is easy to appreciate that those populations that are wealthier and better educated employ much larger quantities of stocks in the service of a better quality of life than do those less well off.

5.3 Determining In-Use Stocks

In principle, it is straightforward to determine the level of in-use stocks. There are two possible methods of doing so. One is termed the "bottom-up" method and is achieved as follows:

- List the principal reservoirs of stocks of materials (buildings, transport, communication facilities, etc.).
- Determine in some way the typical stocks of materials of interest in those reservoirs (by consulting engineering documents, public records, etc.).
- Quantify the number of each of the typical stocks in the region of interest (by observation, tax records, etc.).

A challenge with this approach is that it is inherently incomplete (all reservoirs cannot be surveyed) and that substantial on-the-ground effort is required. Consider, for example, a bottom-up analysis of in-use stocks of nickel (an important metal with many technological uses) in the city of New Haven, Connecticut, USA (Rostkowski *et al.*, 2007). The authors first determined the major societal sectors within which nickel was likely to be used — they identified six: buildings and infrastructure, transportation, industrial machinery, household appliances and electronics, metal goods, and other end use. For each of these sectors, the researchers then counted all of the different nickel-containing products and the amount of nickel in each, summing to the total for the city and per capita for each resident. Their results for one of the sectors appear in Table 5.1.

The second general method of stock determination, the "top-down" approach, is accomplished by analyzing overall flows for the material of interest (Chapter 6 will cover the technique of material flow analysis in

Table 5.1. Weight of nickel in metal goods and other end uses in New Haven (Rostkowski *et al.*, 2007).

End use	Ni in use (Mg)	Ni in use per capita (g)
Domestic equipment	66.7	539
Tableware	1.3	11
Cookware	59.5	481
Tools	3.3	27
Household batteries	0.6	5
Disposable razors	0.1	0
Gardening tools	1.9	15
Commercial equipment	15.0	121
Kegs	10.0	81
Kettles	0.1	1
Tables/stations/bars	2.3	19
Tableware	0.1	0
Cash register drawers	2.4	20
Restaurant industry	15.9	129
Fasteners	9.3	75
Jewelry	0.4	3
Coins	5.0	41
Hospital equipment	1.6	13
Total	113.9	921

detail). This approach requires estimations for a sequence of time intervals (usually yearly), in which input and output flows are quantified. Top-down approaches tend to be possible where measurements of appropriate flows are routinely available, a situation that favors countries with good import-export flows, accurate vehicle registrations, careful mining and processing statistics, etc. Given this information, one proceeds as follows. Assume one wishes to determine the stock σ of material m in a repository, such as steel in infrastructure. During a specified time period, the repository receives an inflow I and loses material in an outflow O. This gain–loss sequence is repeated a number of times, often over the years for which material flow data are available. The in-use stock at the end of the time sequence of *n* years is then given by

$$\sigma_m = \sum_{y=1}^{n} \left(I_{m(y)} - O_{m(y)} \right) \tag{5.1}$$

where σ_m is the stock of material, $I_{m(y)}$ is the inflow of material *m* during year *y*, and $O_{m(y)}$ is the outflow of material *m* during year *y*.

A significant limitation to the reliability of the top-down determination of stocks is that the results depend on a reasonably accurate knowledge of the average lifetime of the product (building, vehicle, appliance, etc.) that contains the material of interest. In some cases, and in some regions, this information can be quite precise (vehicle registration data, for example). It is generally unusual, however, for detailed public or corporate records of most lifetimes of interest to be available. As a consequence, most stock studies are forced to deal with limited data, perhaps drawn from different epochs and different regions and assumed to be "accurate enough".

In general, national statistical agencies do not concern themselves with counting materials. As a result, no extensive compilation of available stock data for different repositories and different regions is currently available and it is challenging to keep individual assessments current. Building lifetimes appears to present a larger challenge than automobiles; they clearly range from a few years to more than a hundred. The lifetimes of infrastructure assets (bridges, ports, etc.) can vary from ten years to centuries depending on the product under study. Overall, it appears that uncertainties in final product lifetimes pose one of the two greatest challenges, the other being uncertainty in the materials content of the products themselves. All of this is not to say that determinations of in-use stocks are without value but that their precision is likely to be moderate at best.

5.4 The Utility of Stocks Information

The results of in-use stock determinations provide insight into the ways in which material stocks develop over time and on the material

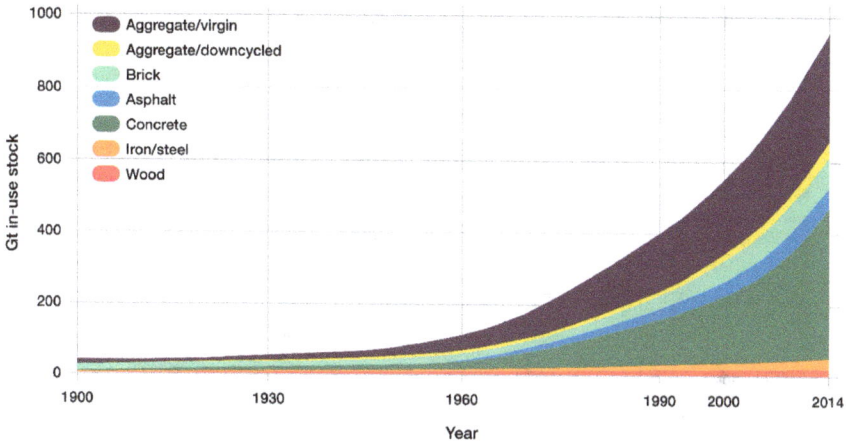

Figure 5.3. Material stocks in use (Wiedenhofer *et al.*, 2019).

compositions of stocks. In an influential global stock dynamics study, Wiedenhofer *et al.* (2019) estimated the mass of in-use stocks from the year 1900 to the present. The result is shown in Figure 5.3. It is immediately apparent that concrete constitutes half of all stocks, aggregate (crushed stone) perhaps another 30%, and brick 5–10%. Therefore, all other materials — plastics, paper, wood, metals, glass, and so forth — together make up only 10–15% of all in-use stock. Because all stock exists to serve human purposes, it is easy to conclude that the principal uses of concrete, aggregate, and bricks are vitally important to human well-being (in the forms, of course, of buildings, roadways, and simpler forms of infrastructure). Of course, plastics, paper, wood, metals, glass, and other materials have important functions to play in providing services that we desire, but they do so in amounts much smaller in mass than those of the more traditional materials.

Less broad but perhaps equally useful can be more detailed stock analyses of specific materials or material groups. In an analysis of chromium in Japanese steel (Oda *et al.*, 2010) the authors derived year-by-year flows of alloy steels, subdivided into six categories. The cycle for 2005 is shown in Figure 5.4(a). A comparison of input and output flows indicates that there are net additions to in-use stock in nearly every alloy

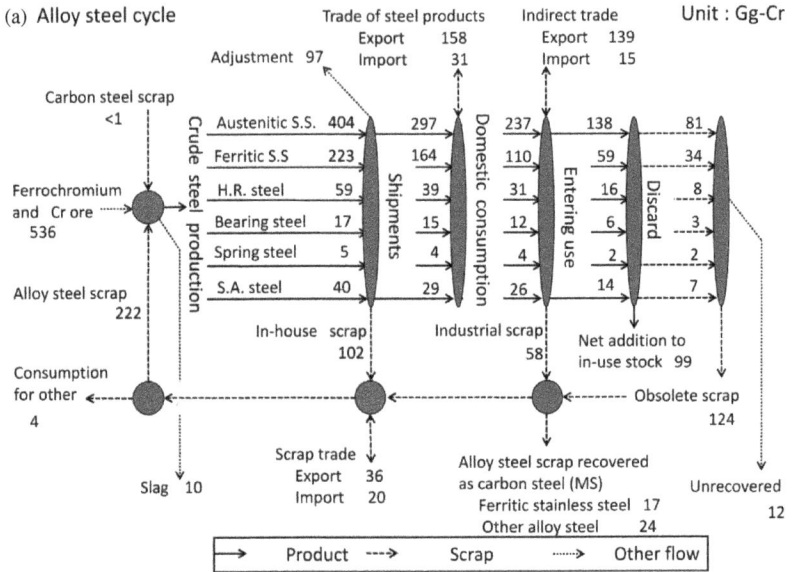

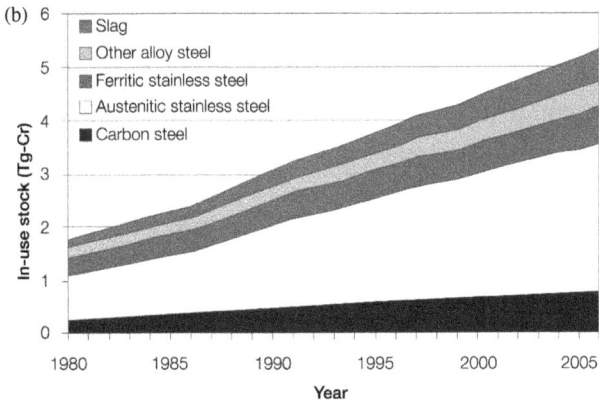

Figure 5.4. (a) The flows of chromium (Gg Cr/yr) and (b) additions to in-use chromium stocks (Gg/Cr) in Japan in 2005 (Oda *et al.*, 2010).

steel category, with austenitic stainless steel leading the way. When the chromium in all steel categories is taken into consideration in a year-by-year analysis, the result is a graph over time of the chromium in-use stocks (Figure 5.4(b)). Over the 25-year period of the analysis, chromium stocks more than doubled. Such information is quite important to steel

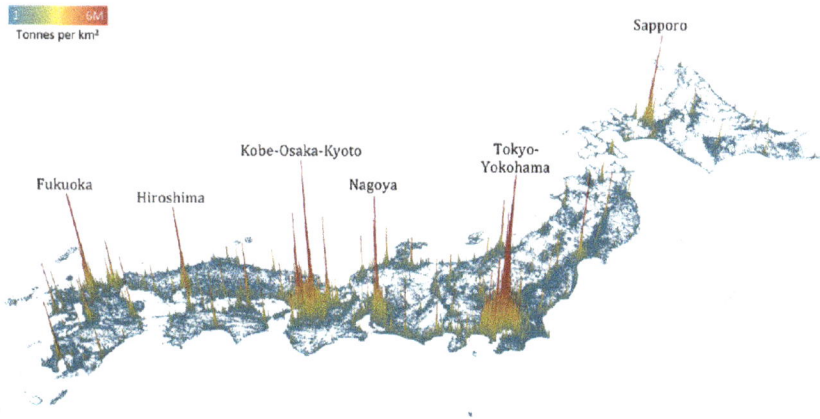

Figure 5.5. The spatial depiction of steel stock in Japan (Tanikawa *et al.*, 2015).

recyclers, as successful separation and recovery of chromium-containing steels when they come out of use are much more profitable than recycling them as regular carbon steels.

In-use stocks may also be determined on a spatial basis. In a study that involved surveys both by images from space and from on-the-ground studies, Tanikawa and colleagues generated a Japan-wide study of the total mass of material in major buildings. The results were presented in a dramatic diagram reproduced here as Figure 5.5. The metropolitan conglomerate of Tokyo-Yokohama and other major urban areas are very easy to distinguish, even without the labels provided by the diagram. Several years later, when a tsunami off the Japanese coast caused major damage, especially in northern Japan, the stock information from the earlier study enabled accurate and rapid provisioning of the materials needed to rebuild those parts of the country that had suffered major damage.

5.5 Policy Aspects of Stock Analyses

In-use stocks reveal a number of different aspects about the long-term use of materials, the development of urban areas, and the prospects for technologies. Several examples of how this industrial ecology concept can be both interesting and useful in urban planning and technology planning are given below.

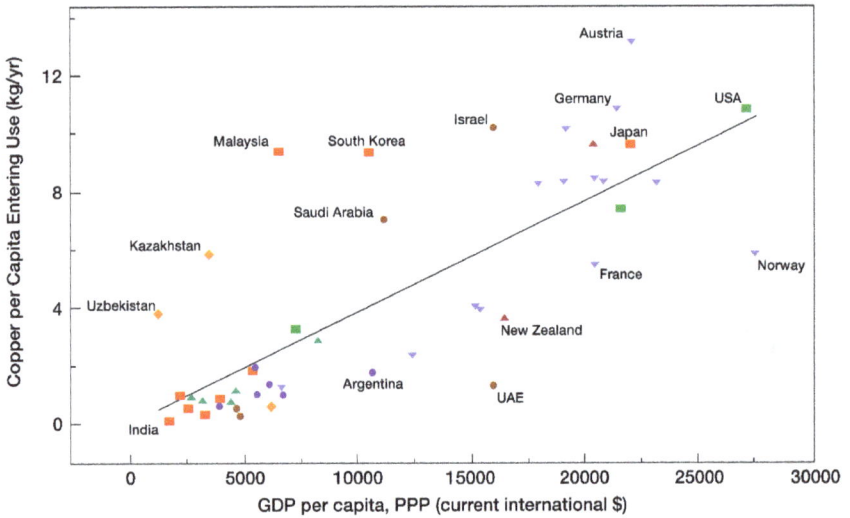

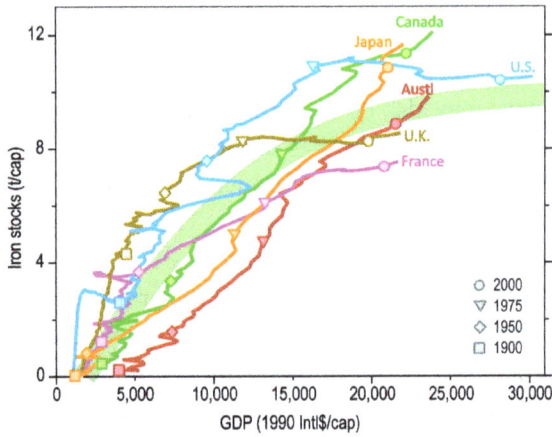

Figure 5.6.   (a) Copper stock as a function of GDP/capita for different countries (adapted from Binder *et al.*, 2006). (b) The per-capita saturation of iron in-use stock (Müller *et al.*, 2011).

Two parametric evaluations that have proven of significant interest deal with resource use as a function of wealth. Figure 5.6(a) shows that per capita copper stock demand is a roughly linear function of per capita income, although there is lots of scatter around the best fit line, with South Korea and other countries substantially higher and Norway and other countries substantially lower. Nonetheless, the correlation provides a rough estimate of copper demand as countries improve the average income of their citizens. The second study repeated the concept of Figure 5.6(a) but for iron in-use stock rather than annual copper flow into use. The result showed that some countries, including the United States, the United Kingdom, and France had not only increased the per capita stock over time but had also reached per capita saturation of iron stock at a wealth level of 20,000–25,000 international dollars per capita Figure 5.6(b). Presumably, these countries would only require iron in the future to accommodate population growth or to maintain existing facilities, a result that provides long-term insight on iron demand.

Although in-use stocks are generally beneficial, they can inhibit change even if change might be desirable. An example of this conundrum is the transition from fossil fuel energy to renewable power sources. From an environmental standpoint, such a change is desirable. However, it involves new investment, worker retraining, and abandoning the physical in-use stocks that make up fossil fuel infrastructure, or finding other uses for them, and it may be simpler and cheaper in the short term to leave the current stocks in place. If material stocks are not already in place, however, transitions from the absence of a desirable stock to its ubiquitous presence can be rapid, as shown by the growth of solar photovoltaic installations in areas without fossil fuel infrastructure.

Further Reading

Bar-On, Y.M., R. Phillips, and R. Milo, The biomass distribution on Earth, *Proceedings of the National Academy of Sciences of the United States, 115*, 6506–6511, 2018.

Binder, C.R., T.E. Graedel, and B. Reck, Explanatory variables for per capita stocks and flows of copper and zinc, *Journal of Industrial Ecology, 10*(1–2), 111–132, 2006.

Cabrera Sorrenho, A., and J.M. Allwood, Material stock demographics: Cars in Great Britain, *Environmental Science & Technology, 50*, 3002–3009, 2016.

Gerst, M.D., and T.E. Graedel, In-use stocks of metals: Status and implications, *Environmental Science & Technology, 42*, 7038–7045, 2008.

Lin, C., G. Liu, and D.B. Müller, Characterizing the role of built environment stocks in human development and emission growth, *Resources, Conservation, and Recycling, 123*, 67–72, 2017.

Müller, D.B., T. Wang, and B. Duval, Patterns of iron use in societal evolution, *Environmental Science & Technology*, *45*, 182–188, 2011.

Oda, T., I. Daigo, Y. Matsuno, and Y. Adachi, Substance flow and stock of chromium associated with cyclic use of steel in Japan, *ISIJ International*, *50*, 314–323, 2010.

Pauliuk, S., R.L. Milford, D.B. Müller, and J.M. Allwood, The steel scrap age, *Environmental Science & Technology*, *47*, 3448–3454, 208.

Rostkowski, K., *et al.*, "Bottom-up" study of in-use nickel stocks in New Haven, CT, Resources, *Conservation, and Recycling*, *50*, 58–70, 2007.

Tanikawa, H., T. Fishman, K. Okuda, and K. Sugimoto, The weight of society over time and space: A comprehensive account of the construction material stock of Japan, 1945–2010, *Journal of Industrial Ecology*, *19*(5), 778–791, 2015.

Wiedenhofer, D., T. Fishman, C. Lauk, W. Haas, and F. Krausmann, Integrating material stock dynamics into economy-wide material flow accounting: Concepts, modelling, and global application for 1900–2050, *Ecological Economics*, *156*, 121–83, 2019.

Zhang, L., Z. Yuan, and J. Bi, Estimation of copper in-use stocks in Nanjing, China, *Journal of Industrial Ecology*, *16*, 191–202, 2012.

Chapter 6

Material Flow Analysis

Chapter Concepts

- Material flow analysis (MFA) is the methodology that is used to construct a socio-economic metabolism for a chosen industrial material.
- MFAs can be generated at different spatial levels and organizational types.
- Dynamic MFAs construct an industrial metabolism over time, thereby revealing in-use material stock dynamics.

6.1 Introduction

Material flow analysis (MFA) is an analytical response to the desire to study modern society in physical, quantitative detail, with the realization that a great many aspects of that society — housing, food, transport, medicine, and so forth — are built on the backs of materials. Until the early 21st century, however, little quantitative information was available concerning rates of material use, material losses to the environment, efficiency of recycling, and other quite basic information about human use of materials. Lord Rayleigh famously said in 1883 that "when you can measure what you are speaking about, and express it in numbers, you know something about it; when you cannot express it in numbers your knowledge is of a meager and unsatisfactory kind." MFA has evolved to provide satisfactory information related to material flows and stocks in many instances.

MFA is one of the central methodologies of industrial ecology. It is through MFA that an "industrial metabolism" (the flows of resources into and from a particular entity of human society) can be mapped and quantified, much as an accountant determines and quantifies monetary deposits and withdrawals into and from a bank account. Dynamic MFAs (those that treat a specific region or system over time) go further; they permit a determination of the in-use and "hibernating" stocks of materials in an industry or society (the material version of the accountant's "assets and liabilities").

Unlike the accountant, however, who deals only with stocks and flows reported in monetary terms, the MFA analyst faces a wide diversity of commodities — biomass, polymers, metals, minerals, and more, and transactions that often deal with inadequately described categories (e.g., "iron and aluminum alloys"), lumped categories (e.g., "plastics"), or resource flows that are seldom or never measured (many of the discard flows). MFA-related information quality may vary, from carefully measured data to rough estimates to conjecture. MFA also needs to address flows that are of little import to the accountant because they are not monetized, such as waste flows not captured or emissions to the environment. The MFA specialist must therefore be part detective, part archivist, part extractor of information from experts, and part bold estimator, in order to build the internally consistent database needed to achieve a useful MFA.

In principle, MFA approaches can be applied to any material or combination of materials. In practice, metal stocks and flows have thus far proven to be particularly suitable for analysis, largely because they can often be tracked relatively easily and because data are commonly available for at least some parts of their life cycles. Moreover, MFAs can also treat groups of materials, such as construction minerals (sand, crushed stone, cement) or summed material flows into and from a country or region.

6.2 Flows, Stocks, and MFA Methodology

Industrial ecology is a concept with cycles at its very heart, and cycles are analyzed by means of materials and energy budgeting. Nearly

Figure 6.1. A simple conceptual system for material flow calculations. The water level in the tub over time is determined by the water flows in and out during each period.

everyone is familiar with the concept of a household or personal financial budget, whether or not she or he is conscientious about making and sticking to it. An approach very similar to that of financial budgeting is used to fashion budgets in industrial ecology. The situation can be appreciated with the aid of the diagram in Figure 6.1, which shows a tub receiving water from a faucet and having a drain of specific size. When the water is supplied at a constant rate by the faucet and is removed at an equal total rate by the drain, the water level remains constant. When the tank has multiple faucets and drains, however, and has some wave motion that makes it difficult to tell whether the absolute level is changing, an observer may not be able to know whether or not the system is in balance. In that case, they may try instead to measure the rate of supply from each of the faucets and the rate of removal in each of the drains over a period of time to see whether the sums are equivalent. A part of this technique involves the determination of the pool size (the total quantity of water in the tank) and either the rate of supply or the rate of removal. Determination of changes in the pool size then gives information about rates that are difficult to measure. The process of estimating or measuring the input and output flows and checking the overall balance by measuring the amount present in the reservoir constitutes the budget analysis.

Suppose that the input from one of multiple sources is increased, i.e., in our analogy, the flow from one of the faucets increases. Will the water level keep increasing? The answer depends on whether one of the drains can accommodate the additional supply. If no such drain is present, then the water level will indeed increase. Conversely, if the flow into a drain is enhanced for some reason, such as the removal of an obstruction, then the water level will decrease.

All budgets involve the same concepts mentioned above. One is that of the reservoir or *stock*, in which material is stored. Examples include the shipping department where completed products are prepared for forwarding to customers, or the atmosphere where emissions of industrial vapors collect and react. A second concept is that of flux or *flow*, which is the amount of a specific material entering or leaving a reservoir per unit of time. Examples include the rate of evaporation of water from a power plant cooling tower or the rate of transfer of ozone from the stratosphere to the troposphere. Third, we have sources and sinks, which are rates of input and loss of a specific material to or from a specific reservoir per unit of time. A system of connected reservoirs that transfer and conserve a specific material is termed a cycle. Industrial ecology budgets follow this pattern: determination of the present stock level (the concentration of a single material or a group of materials), a measurement or estimate of sources, and a measurement or estimate of sinks. A perfect determination of any two of these three components determines the other, as a consequence of the *conservation of mass* principle: material can be transformed, but not lost. Because any material of interest in an industrial facility or in the environment may have several sources and sinks, each one must generally be studied individually.

The MFA process begins with defining the system to be studied and constructing a diagram of the material flows and stocks that occur within that system. This important first step requires detailed knowledge of the system: the points in the system at which material is designed to enter and leave, the points at which material can be stored for shorter or longer periods of time, the nature and location of the transformations that occur, and the locations at which waste products are generated and exported outside the system boundary. Unless the system is carefully and properly specified, the necessary data will not be identified nor acquired, the necessary data may not be pursued, and the resulting system specifications may be incomplete, inaccurate, or of little ongoing use.

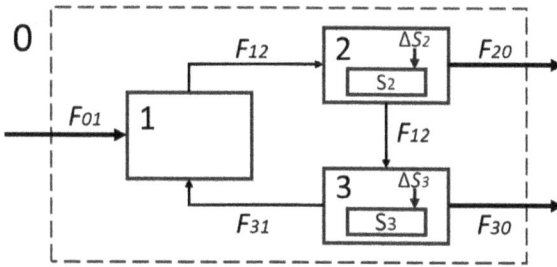

Figure 6.2.   (a) A process and system balance diagram for a simple system. (b) A performance indicator diagram for the system in (a).

A typical schematic diagram for industrial ecology material flows is pictured in Figure 6.2. The system boundary specifies what is to be studied (the element or other material), for what spatial entity (often a country or region), and for what time interval (often a year). The individual boxes are termed "processes"; they are identified by numbers or labels and the material flows among them are shown by arrows. Each process may involve an internal stock (the small box within the larger one) and a flow into or from that interior box (the net addition to or subtraction from stock). This diagram therefore describes a system that incorporates three processes that exchange and perhaps store material, as well as flows across the system boundary to/from an external reservoir (labeled "0").

The system can be defined mathematically through a set of equations. First, the system variables can be specified:

$$\text{Stocks: } S_2(t), S_3(t)$$
$$\text{Flows: } F_{01}(t), F_{12}(t), F_{23}(t), F_{20}(t), F_{30}(t), F_{30}(t)$$
$$\text{Net additions to stock: } \Delta S_2(t), \Delta S_3(t)$$

Then writing a mass balance equation for each individual process, and the system as a whole, gives:

$$\text{Process 1: } F_{01}(t) + F_{31}(t) - F_{12}(t) = 0$$
$$\text{Process 2: } F_{12}(t) - F_{23}(t) - F_{20}(t) = \Delta S_2(t)$$
$$\text{Process 3: } F_{23}(t) - F_{31}(t) - F_{30}(t) = \Delta S_3(t)$$
$$\text{System: } F_{01}(t) - F_{20}(t) - F_{30}(t) = \Delta S_2(t) + \Delta S_3(t)$$

Given the parameters thus far defined, various indicators of performance can now be derived:

Indicator #1: Efficiency = useful output (t)/total input (t)

$$\text{Process 1: } \varepsilon_1(t) = F_{12}(t)/F_{01}(t)$$
$$\text{Process 2: } \varepsilon_2(t) = F_{20}(t)/F_{12}(t)$$
$$\text{System: } \varepsilon_s(t) = F_{20}(t)/F_{01}(t)$$

Indicator #2: Emissions intensity = emissions/useful output

$$\text{Process 2: } \varepsilon_2(t) = F_{20}(t)/F_{23}(t)$$
$$\text{Process 3: } \varepsilon_3(t) = F_{30}(t)/F_{12}(t)$$
$$\text{System: } \varepsilon_s(t) = F_{20}(t)/F_{31}(t)$$

This MFA vision and methodology can be summarized by a list of attributes customarily possessed by any resulting analysis:

 (i) An MFA is the study of a clearly designed material flow *system*, not merely the study of a particular material flow.
(ii) An MFA includes a description of each flow in the system (e.g., the physical and chemical state of each material).
(iii) An MFA quantifies all flows of significance in the system. Conservation of mass constraints apply at each of the system nodes.
(iv) The presentation of MFA results is generally diagrammatic as well as numeric.
 (v) An MFA analysis includes a discussion (or, better yet, a detailed analysis) of the reliability of the results.

No matter what the cycle, the careful construction of a flow diagram is essential if an accurate picture is to be obtained.

6.3 Cycles and Data in MFA

An industrial ecology resource analysis may, in principle, deal with any spatial scale, though global and country-level MFAs are by far the most common. The effort may also deal with many reservoirs, although something like a half-dozen is probably most common. A generic metal cycle,

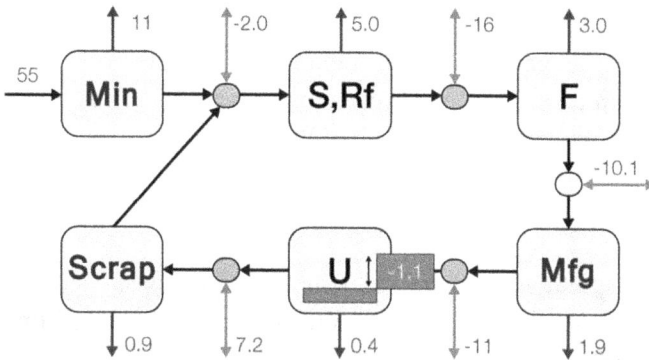

Figure 6.3. An example material flow analysis for a metal. Min = mining, S = smelting, Rf = refining, F = fabrication (of intermediate products), Mf = manufacture (of final products); U = use, Rc = recycling; circles denote markets for each type of metal-containing product. All flows are measured in convenient metal mass units, such as Mg/yr.

shown in Figure 6.3, is an example. There are six (or seven, depending on how the cycle is constructed) sequential processes (or reservoirs or life stages), linked by arrows. First the ore (rock containing metal-rich minerals) is mined and the minerals separated from the rock, followed by Smelting (to separate the metal minerals from the rock debris) and Refining (to purify the nickel) [the Smelting and Refining steps are sometimes combined]. At the Fabrication stage, the metal is fashioned into intermediate products such as rolls and sheets, which are then employed in the Manufacture of products such as appliances or vehicles. Products are employed in Use (generally for a number of years), and then are discarded to Recycling, where they may be recovered for reuse, sold as scrap (presumably for reuse), or lost. No process is perfect, however, and losses at each stage are shown as gray arrows.

The cycle in Figure 6.3 is characterized by processes that are linked through "markets" (the intermediate circles), each indicating trade with other regions (light gray arrows) at the respective life stages, and losses to the environment (dark gray arrows). The scrap market plays a central role in that it connects Waste Management with Refining and Fabrication. The Use process incorporates a stock repository that may increase or decrease (black arrow). The cycle is surrounded by entities lying outside the system boundary: trade partners (other regions), the lithosphere from which ore extraction takes place, and repositories for nickel in tailings and slag (i.e., production wastes), and in landfilling.

Example 6.1

Imagine that you are assigned to quantify the flows and stock of a particular material whose system is pictured in Figure 6.3. All import and export flows are measured (in unspecified units), as is the flow leaving the in-use stock. Compute the missing flows.

Solution:
Process Min: 55–11 = 44; Process S,Rf: 44+7–2–5 = 44;
      Process F: 28–3 = 25
Process Mf: 25–10.1–1.9 = 13; Process U: 13–11–0.2–1.1 = 0.7;
      Process Rc: 0.7+7.2–0.9 = 7.0

Material exports (negative flows from the system under analysis, i.e., sales) indicate a manufacturing process that is producing and selling products. Imports of material occur only in the recycling stage (except the original input) and are used to minimize raw material extraction.

A frequent challenge in MFA is getting an accurate assessment of the flow from in-use stock. Typically, the stock of a particular element will be distributed among several types of uses, each of which has its own in-use lifetime. An element contained in home appliances will typically be in use for two or three decades, while the same element in a smart phone may leave use in a year or two. This fact introduces a variable time delay into the system, meaning that material often becomes available for urban mining a decade or more after its initial entry into the system. This delay must be considered in any efforts to quantify the amount of material likely to be available for reuse, and the forms in which it will be discarded. Product lifetimes are not routinely measured, and thus need to be determined on the basis of expert judgment, field surveys, or other means. Such estimates often are the only way to proceed. In other cases, however, reasonably good data may be available, as in the following example.

Example 6.2

Assume that element E has four principal uses in a country, and that the flows into use are well-determined (as sometimes happens if we're lucky). What is the total outflow of E after several years of use?

Over a 6-year period the flows are:

	Flows into use (Mg E/yr)					
Use	2015	2016	2017	2018	2019	2020
1	45	60	70	60	75	80
2	30	35	40	30	40	50
3	15	17	20	15	18	25
4	10	10	12	15	16	20

Solution:

For use 1 with lifetime 4 years, the amount of E released from in-use stock is that which entered in 2017 (4 years earlier): 70 Mg E. Similarly, Use 2s flow from use is that which entered in 2015: 30 Mg E, for Use 3: inflow in 2018 = 15 Mg E, and for Use 4: 20 Mg E.

The total outflow in 2021 for all of element E's major uses is thus $70 + 30 + 15 + 20 = 135$ Mg E.

All of this MFA seems like a lot of effort. Why do we wish to do this? There are many reasons for wanting to quantify these flows, including:

- Manufacturers wish to know how their operations connect with other product life stages.
- Market analysts search for this information in order to become better informed.
- Taxes, tariffs, and other fees are paid on the amounts and types of imports and exports.
- Environmental scientists are interested in rates of material loss (and related impacts).
- Sustainability researchers want to know the effectiveness of recycling.

Where does the information for generating a MFA come from? Some examples are:

- Government reports on material use, use sectors, and loss. (The United States Geological Survey's Mineral Commodity Summaries is widely used, for example.)
- Import/export data. The UN Comtrade database is widely employed for such information, although its level of detail may in some cases be problematic.
- Corporate reports. Corporations make public selected material flow information in annual reports, conference presentations, speeches, and the like.
- Market analysts. A number of private companies collect and synthesize data and comments from corporations and other experts.
- Market surveys. (e.g., what percent of a specific population discarded personal electronics in the last calendar year?)
- Informed estimates. It is often possible to informally ask industry experts specific questions about flows and markets of interest for an MFA analysis.

How can one be sure that stocks or flows with the same name but reported by different sources are really reporting on the same flow?

- Various national and international standards can be useful in this regard if they are carefully explained in related documents, although poorly-described data remain a problem.

None of the above suggestions promise surefire success in procuring the desired information. Why might data be missing or not useful?

- Corporate or government secrecy;
- The data are not disaggregated enough to be useful;
- Nobody has measured the stocks and/or flows.

How can missing data be dealt with? Among the many possibilities are:

- Employ the principle of conservation of mass. (If information for all but one flow into a node is available and you believe no significant in-use stock change is occurring, the missing flow can be approximated by addition and subtraction.)

- Use data from other countries, regions, etc. as proxies.
- Restrict the scope of the analysis. (A common approach is to determine flows up to and including in-use reservoirs but to not address in-use lifetimes and recycling flows.)
- Restrict the scale of the analysis. (For example, produce a European-level analysis instead of a country-level analysis. Global data are often more available and more accurate than regional data, which in turn may be more accurate than country-level data.)
- Infer flows of your target material from related information. (For example, alloy flow information [e.g., tool steel] provides information about the individual elements in the alloy [e.g., tungsten and cobalt in tool steels].)

An example of a carefully prepared MFA analysis is shown in Figure 6.4, a flow diagram for the zinc cycle of Asia in 2010. Several features are easy to deduce, including the largest input flow (yellow arrow, from domestic Chinese mines), the large reuse of scrap (blue), and the large rate of loss at end of use (green). Trade with other regions occurs between life stages at "markets" (the small black circles between the flows). A small dashed-line insert shows input to the Scrap node to reflect an input flow that is required to balance input and output, but for which available data do not determine definitive assignment to one of the larger flows.

An extension of a static MFA analysis is a "dynamic MFA" that consists of a sequence of cycles, usually year by year for a decade or more. A benefit of such an analysis is that it provides a quantitative estimate of in-use stock — the cumulative difference of flows into and out of use. Examples of in-use stock determinations utilizing MFA approaches are illustrated in Chapter 5.

It is common that the data on material flows, having been collected from many different sources, are not completely consistent. This challenge can be dealt with in two ways. The first is to make one's best assumption as to which of the variables is more likely to be correct and to adjust the other(s) to achieve numeric consistency. The second is to admit that the problem exists and to call attention to it as a need for future data collection. Calling attention to an MFA data problem can be done in various ways, but perhaps the most common is to add a distinctive flow and

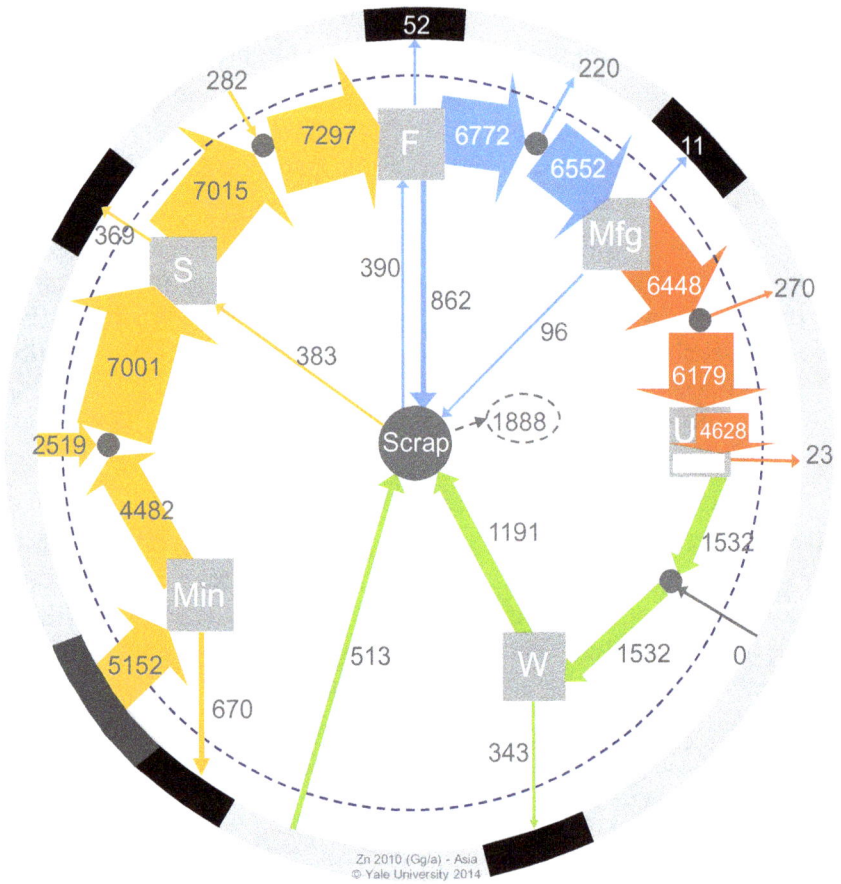

Figure 6.4. The 2010 Zn cycle for Asia. The line widths are proportional to the magnitudes of the zinc flows from one node of the diagram to the next. The colors indicate flows of zinc during ore processing (yellow), fabrication (blue), manufacturing (tan), and discard, recycling, and loss (green). Min = mining, S = smelting, F = fabrication, Mfg = manufacturing, U = use, W = waste management. The units are Gg/yr. (Meylan and Reck, 2017).

reservoir to the diagram, as shown by the dashed line oval from the Scrap sector in Figure 6.4 to complete the resource balance but to indicate where the discrepancy occurs.

In some cases, as in Figure 6.4, the existence of flows crossing a boundary may be well known, and the principal challenge is to quantify them (and the related stocks as well). In others, as in atmospheric chlorofluorocarbons, the importance of some of the sources and sinks may not be known and must be deduced by quantifying those flows for which information is available. In all cases, however, the process begins by specifying the metabolic cycle, its intermediate forms, and its processes in as much detail as possible.

A high level of detail is desirable in an MFA. The additional work needed to generate such a product is substantial, but the result can have considerable value. Such an achievement is shown in Figure 6.5 — the global cycle of the important nutrient phosphorus. Inputs, flows, and losses are clearly shown, and the opportunities for flow intercept are clearly visible.

In industrial ecology, the concepts of budgets and cycles are applied not only to the anthropogenic use of resources but also sometimes to the combination of anthropogenic and natural processes. The results help evaluate present metabolic needs and to estimate those that may be required in the future. Similarly, it may be possible to study specific resources as they pass through various technological organisms, and thus evaluate resource supply, use, and loss.

6.4 Scope of Materials Being Addressed in MFA

As the protocols and approaches of MFA became reasonably well-structured in the early part of the 21st century, the applications were largely to three metals: iron, aluminum, and copper. In decades since that time a large number of other elements and materials have been studied by MFA approaches.

Petrochemicals and plastics differ from metals in that they cannot readily be recovered, reprocessed, and reused. The traditional end-of-life approach has been either to burn them for energy production or to down-cycle them into products such as park benches. Material flow data for these materials were published in peer-reviewed journals as early as 1998. These MFAs have often been for specific plastics, generally the consumer plastics polyethylene, polypropylene, polystyrene, and polyurethane. A particularly detailed MFA example for eight different polymers is shown in Figure 6.6, which details the initial forms and final uses of the plastics (packaging, construction, electrical, furniture, etc.) and then their disposal and landfilling. Such diagrams provide much insight at a quick glance,

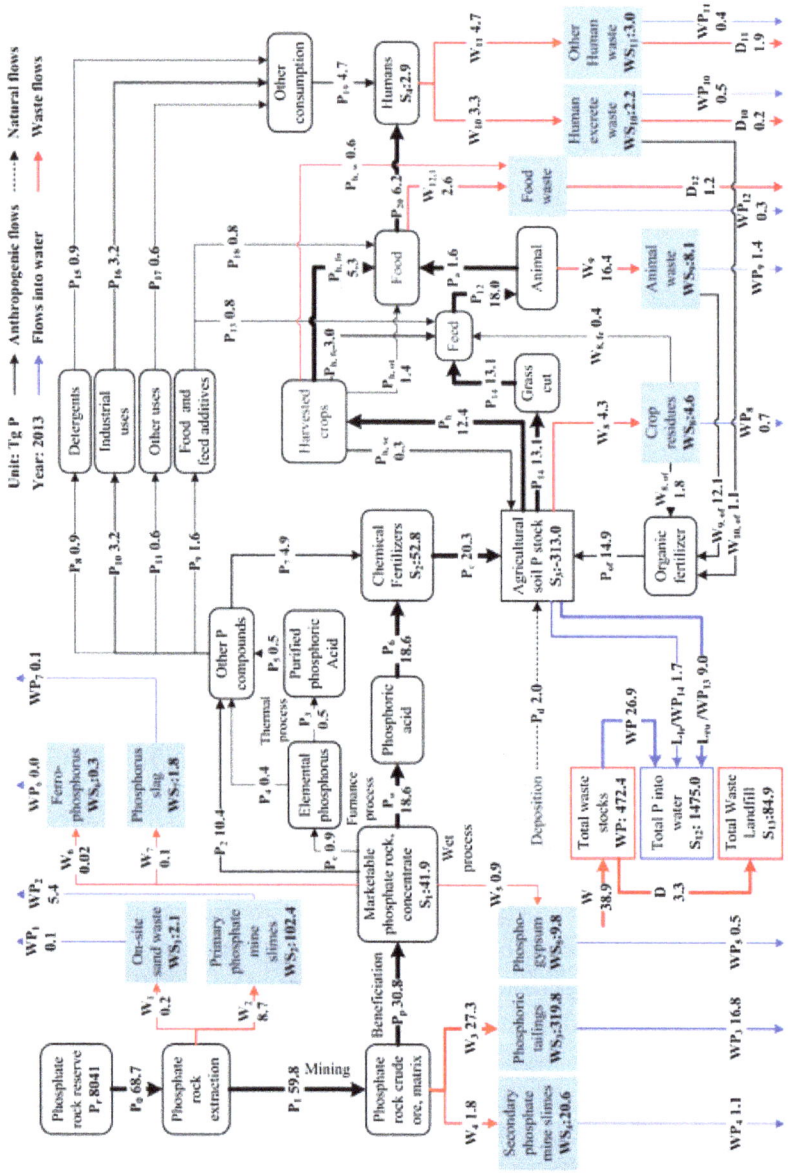

Figure 6.5.   The global cycle of phosphorus in 2015 (Chen and Graedel, 2016).

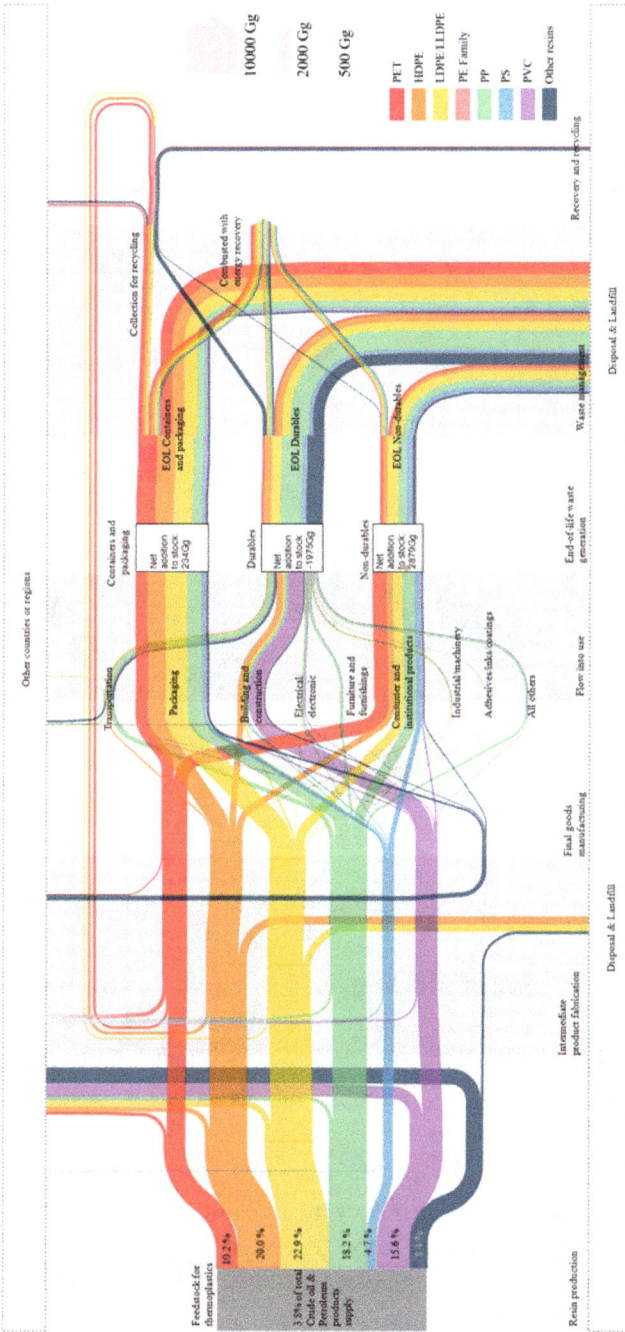

Figure 6.6. Material flow diagram for the predominant plastics manufactured in the US in 2015 (Di *et al.*, 2021).

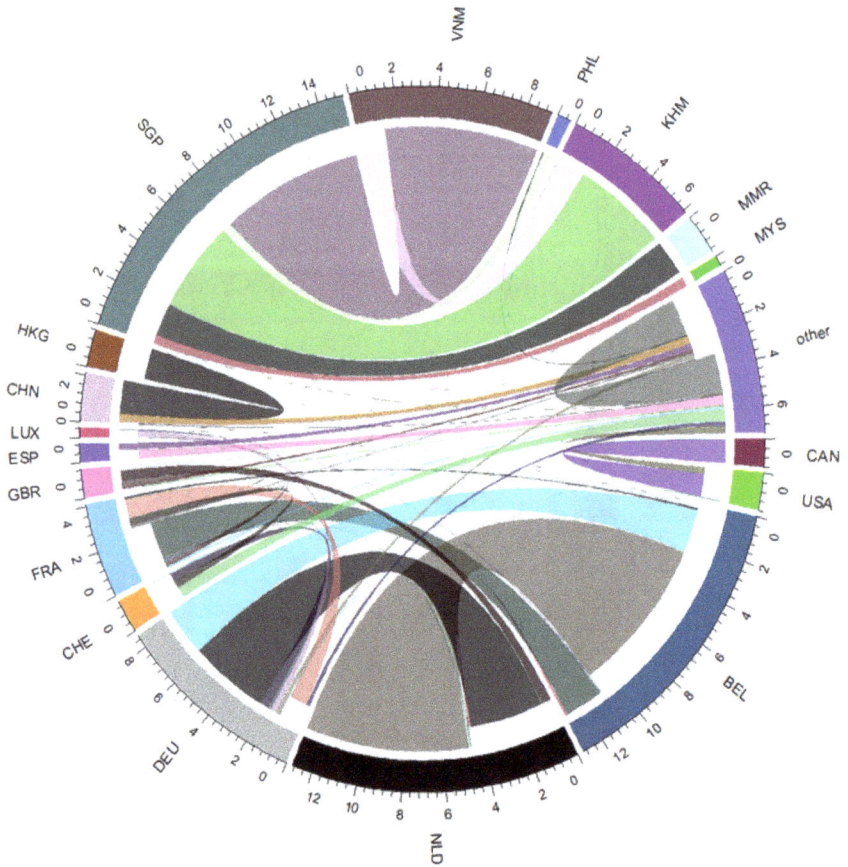

Figure 6.7.　The trade in sand between countries in 2009–2010 (Ioannidou *et al.*, 2020).

such as the importance of packaging as a use as well as the very low recycling flows.

A second example addresses sand, a constituent vital to modern construction. Construction flow amounts are very large — larger even than food or fossil fuels, but most of these materials have low value and are mined and used locally, properties that tend to limit the availability of data. Unlike other minerals, however, data for sand seem well enough established to enable analyses to be generated. Figure 6.7 illustrates sand trade results from country to country in 2009–2010. In this chord diagram the flows connecting the countries are from the exporting country (the external arch) to the importing country (the internal arch). The three-letter codes identify countries and regions. Major exchanges of sand occurred from Cambodia to Singapore, from Belgium to the Netherlands, and from China to Hong Kong.

Material flow analyses are also increasingly common as regards fibers, food, water, energy, and composites, as quantitative information becomes increasing available and as the positive and negative consequences of these flows become increasingly useful for policy decisions by industries, governments, and others.

6.5 Displaying Material Flow Analysis Results

The flows of material in an MFA system can be simply depicted as lines with labels and numbers, as in Figure 6.3. An alternative approach that visually enhances such a diagram is the Sankey diagram originated by Irish engineer Matthew Sankey in 1898. In Sankey diagrams the magnitudes of resource flows are indicated by line widths. Sankey diagrams have become so common in MFA that they have been termed "the visual language of industrial ecology". These diagrams have a unique ability to illustrate many features of an MFA analysis on a single diagram. Color is often utilized to aid in referring to the magnitudes of particular flows, to indicate additional information such as life cycle stages, or to illustrate a particular property of the analysis.

Software to generate Sankey diagrams became routinely available in 2004, when Helmut Rechberger and colleagues at the Technical University of Vienna developed a software program to minimize the effort involved in generating MFA analyses for a variety of systems. The current version of this "STAN" open source software is freely available for download (see Further Reading) and use. It includes features such as data reconciliation and error propagation and also indicates the degree of uncertainty in non-measured flows.

As mentioned above, MFA results are uncertain to various degrees. Some flows are relatively well-determined, such as the import flow of mineral concentrates. Others, such as the flows of materials contained in products, are generated by averaging information on major product types. In some cases, such as losses to landfills, flow rates are determined by difference or by informed estimation. Given reliability variations from one flow to another on the same diagram, an exemplary MFA will attempt to define the levels of uncertainty for each flow. For example, in 2017 Lupton and Allwood (see Further Reading) displayed estimated uncertainty for each flow by varying the color on a Sankey MFA diagram notable for its clarity and beauty.

Acknowledgment

Aspects of the mathematical description of the dynamic MFA system of Figure 6.2 are abstracted from an opensource presentation by Professor Stefan Pauliuk of the University of Freiburg, Germany.

Further Reading

Bertram, M. *et al.*, A regionally-linked, dynamic material flow modelling tool for rolled, extruded, and cast aluminium products, *Resources, Conservation and Recycling*, *125*, 48–60, 2017.

Chen, M., and T.E. Graedel, A half-century of global phosphorus flows, stocks, production, consumption, recycling, and environmental impacts, *Global Environmental Change*, *36*, 139–152, 2016.

Di, J., B.K. Reck, A. Miatto, and T.E. Graedel, United States plastics: Large flows, short lifetimes, and negligible recycling, *Resources, Conservation, and Recycling*, *167*, 105440, 2021.

Eckelman, M.J., and M.R. Chertow, Using material flow analysis to illuminate long-term waste management solutions in Oahu, Hawaii, *Journal of Industrial Ecology*, *13*, 758–774, 2009.

Graedel, T.E., Material flow analysis from origin to evolution, *Environmental Science & Technology*, *53*, 12188–12196, 2019.

Ioannidou, D., G. Sonnemann, and S. Suh, Do we have enough sand for low-carbon infrastructure? *Journal of Industrial Ecology*, *24* (5), 1004–1015, 2020.

Kawecki, D., P.R.W. Schneeder, and B. Nowack, Probabilistic material flows analyses of seven commodity plastics in Europe, *Environmental Science & Technology*, *52*, 9872–9888, 2018.

Lupton, R.C., and J.M. Allwood, Hybrid Sankey diagrams: Visual analysis of multidimensional data for understanding resource use, *Resources, Conservation and Recycling*, *124*, 141–151, 2017.

Meylan, G., and B.K. Reck, The anthropogenic cycle of zinc: Status quo and perspectives, *Resources, Conservation and Recycling*, *123*, 1–10, 2017.

STAN (Substance Flow Analysis) Software for Material Flow Analysis, (https://www.stan2web.net), accessed March 16, 2021.

Chapter 7

Material Reuse and Recycling

Chapter Concepts

- Few products are designed from the start with recycling in mind, which makes recovery and recycling of materials and parts extremely difficult and expensive.
- Functional recycling that preserves the inherent properties of a material is frequently compromised by mixing materials or fixing them together in products leading to contamination, by applications that are inherently dissipative, or by products for which recycling technology has not yet been developed.
- Functional recycling rates for many materials are very low.
- Material waste streams that cannot be functionally recycled may be downcycled to a less valuable use.

7.1 Challenges for Recyclability

Why can't all materials that are incorporated in products of various kinds be reused or recycled when the use of those products is finished? This seemingly obvious inquiry is actually quite complex, but exploring some of the reasons helps explain our current use of materials and how we might improve. Figure 7.1(a) shows different ways that materials can be lost from recycling. In the figure the various applications of a given material are divided into four categories: "in-use dissipated", "currently

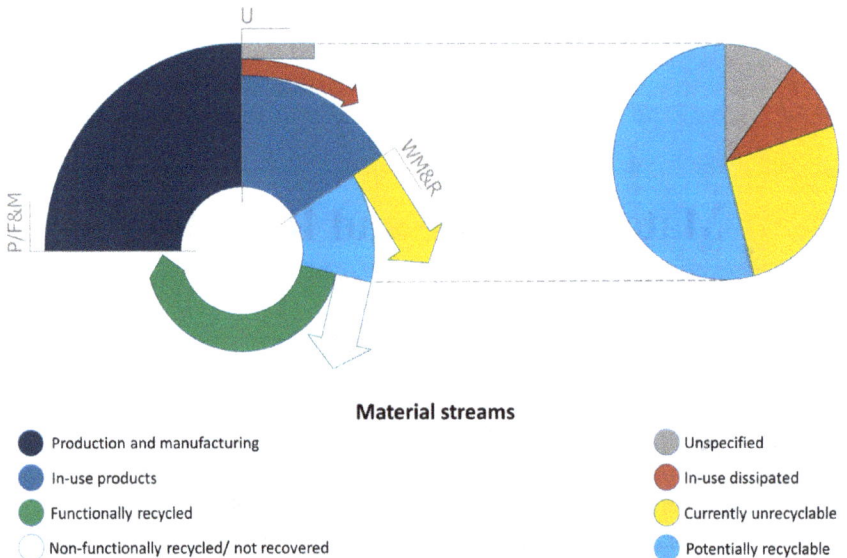

(a)

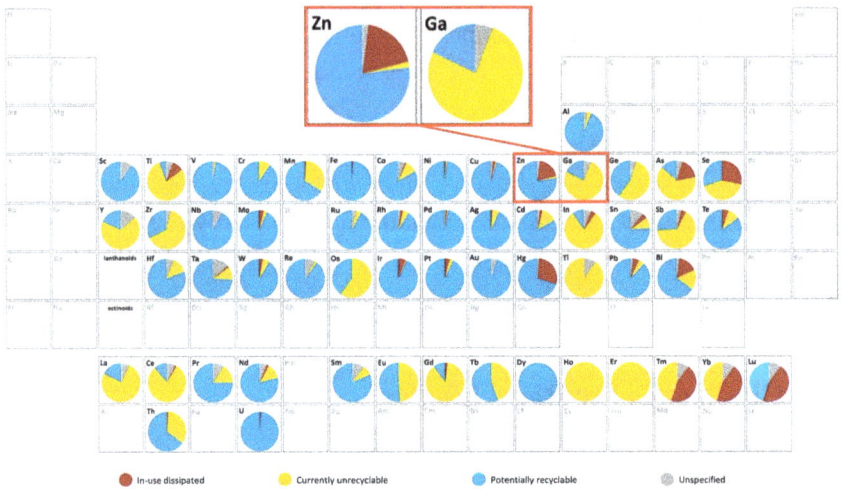

(b)

Figure 7.1.   (a) Material streams for materials in products: dark blue — production and manufacturing; gray — unspecified minor uses; red — in-use dissipated; medium blue — in-use products; light blue — potentially recyclable; yellow — currently unrecyclable; green — functionally recycled; white — non-functionally recycled or not recovered. (b) Percent distributions among the four material streams for a selection of elements (Ciacci *et al.*, 2015).

unrecyclable", "potentially recyclable", and "unspecified" (generally small-scale uses whose low volumes do not justify tracking them). The in-use dissipated category includes material uses that provide immediate benefits (fertilizer ingredients, fireworks, *etc.*) but are designed to dissipate into the environment with little or no prospects for recovery. Some other applications, such as the use of rare earth elements in polishing powders, could be recyclable if a technological recovery and recycling approach had been developed, but are currently unrecyclable as no suitable technologies exist. Finally, in the potentially recyclable category, recycling methods exist but are not employed for reasons of cost, inconvenience, or lack of a sufficient demand. Figure 7.1(b) shows the breakdown among these categories for several elements.

Two examples help to illustrate the challenges of availability for recycling. Consider the diagram for zinc (Zn) (top inset). Nearly two-thirds of all zinc is used as a corrosion protection coating for steel ("galvanized steel"). Over time around 20% of those zinc coatings are dissipated during use; the remainder is typically recovered in steel recycling. Almost all other uses of zinc, such as die casting and alloying, are also recoverable. Gallium (Ga), to zinc's immediate right on the diagram, is mostly used in integrated circuits and other electronic applications. In principle, those products are recoverable and one might assume that a recycling technology has been developed, but no commercially viable alternatives have emerged to date. Some minor uses of gallium in alloys and magnets can, however, be recycled. Similar studies for other elements indicate that recycling potentials that can only be improved by more insightful product design and more extensive development and deployment of new recycling technologies.

In an ideal world, secondary material available through recycling would satisfy the demand for the same material in new products and no virgin resource extraction would be needed. However, materials enter service and remain there in products for extended periods, often decades. This situation is termed the "delaying effect of stocks", a consequence of which is that in a world of increasing demand even perfect recycling is not enough to meet supply (Figure 7.2). At time t_1 the material that entered in-use stocks at time t_0 cannot meet the new higher demand even if completely recycled (recycling rate RR = 100%). If recycling is less than 100% efficient, as is always

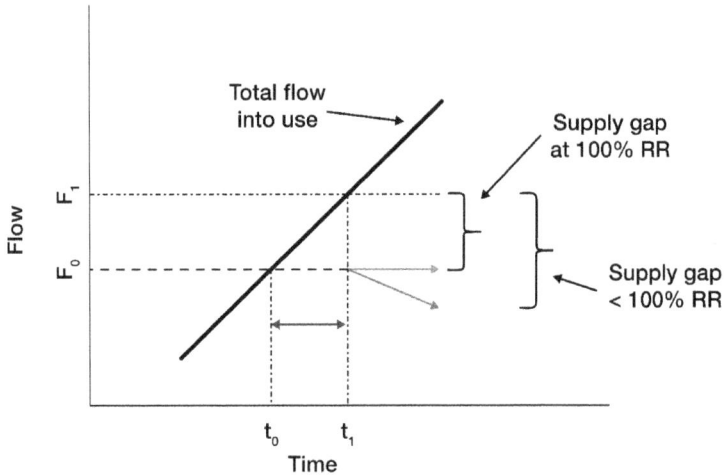

Figure 7.2.   The "delaying effect of material in product stocks in use". In a world of increasing demand even perfect recycling is not enough to meet supply.

the case, the demand/supply gap is even worse. Moreover, some end-of-life products may not immediately be collected for reprocessing or reuse. Personal electronics are famous for being retained in a bedroom drawer for as long as a decade — these are often termed "hibernating stocks". A related category is "comatose stocks" — material that is stored in such a way that it may never be recovered, such as obsolete telephone cables abandoned under city streets. Finally, there are stocks that are designed never to be recovered and reused, such as the foundation pilings under tall buildings and infrastructures; these might be termed "abandoned stocks".

Suppose, however, that a decision has been made to discard a product containing potentially recyclable material. Many steps may be involved in actually carrying out technologically appropriate recycling, as discussed below.

7.2 The Recycling Process

Assume that a material (by itself or contained in a product) is not subject to any of the constraints to reuse and recycling discussed in Section 7.1 and that the product has been discarded. In a successful recycling system, those products are then collected or brought to a recycling facility, separated into

Efficiencies: 50%   x   70%   x   85%   x   95%   =   29%

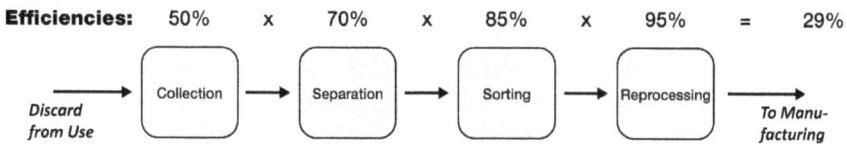

Figure 7.3. The stepwise recycling sequence from discard to newly-available materials. Typical percentage efficiencies at each step are indicated (adapted from diagram developed by C. Hagelüken, Umicore Corporation).

general categories (plastics, metals, etc.), sorted by specific materials (polyethylene, polypropylene, etc.), and sent for reprocessing. This stepwise sequence is shown in Figure 7.3. Separation and sorting steps may be done manually or with a dizzying array of mechanical and optical equipment, but failure to capture components or improper processing occurs to varying degrees all along the sequence. Given the estimated current probabilities for successful processing at each step, the overall efficiency of the overall process usually turns out to be quite low.

Improving this situation requires efforts at all stages of the recycling process, but also in the original product design process. Some of the main strategies of environmental product design are summarized below:

- If possible, capture a product before discard and seek to reuse it elsewhere (this is sometimes termed "relocation").
- If relocation is not feasible, seek to repair or remanufacture the product so as to return it to its original condition and capabilities or, better yet, upgrade or upcycle it to the most recent capabilities of similar products.
- If remanufacturing is not practical, disassemble the product and reuse the components. This step will be enabled by identifying opportunities for efficient identification of the components and researching opportunities for their redeployment. Disassembly is best addressed at the product design stage by minimizing the steps needed for disassembly.
- Components and assemblages that cannot readily be disassembled, or where doing so is not economically or practically feasible, sent on to recycling facilities, where they are separated by sorting and/or shredding, and each of the resulting streams is treated using chemical or metallurgical processes that further separate or purify the material so that it can be used again in new products.

We tend to think of recycling as beneficial, but a challenge arises if a product made in the past contains toxic elements or compounds whose properties were thought sufficiently valuable to offset their toxicity, but must now be managed as a hazardous waste stream. (Cadmium plating of aircraft landing gear as a corrosion preventer is an historic example.) In such cases the current approach is to encase these materials permanently in hazardous waste disposal sites, but Paul Brunner of the Vienna Institute of Technology and his colleagues have long called for designing and implementing "safe final sinks": long-term storage locations where such materials can be safely stockpiled, and perhaps be re-mined in the future if the need arises.

7.3 Quantifying the Recycling of Materials

As a material moves through its life cycle from extraction to manufacturing to use to disposal, there are several discrete pathways for recycling, and recycling rates have been defined in a number of different ways over the years. In 2009 the United Nations International Resource Panel developed a standard definition of flows and a protocol for determining recycling rates, as shown in Figure 7.4(a). The committee made an important distinction between recycled material returned to the scrap market of the element in question [flow (g)], and material still recycled but as a minor component in the recycling of other elements [flow (f)]; the former was labeled "functional recycling" because the physical and chemical properties of the element in question were retained for reuse, and the latter "nonfunctional recycling" because the element, while not lost, no longer was capable of contributing its unique properties to a second use. The "end-of-use functional recycling rate" was then defined as

$$\text{EOL-RR (functional)} = (g)/(d), \qquad (7.1)$$

where flow (d) is the total flow of material in products coming out of use. With this definition, the committee then determined the "best-estimate" functional recycling rate of the elements of the periodic table in one of five percentage ranges, as shown in Figure 7.4(b) (details in Reck and Graedel, 2012). It is easy to see in the figure that 15 to 20 elements have

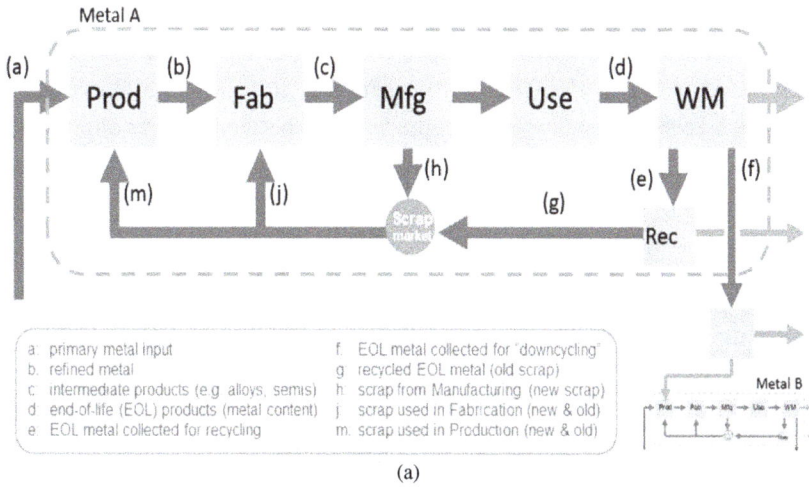

a. primary metal input
b. refined metal
c. intermediate products (e.g. alloys, semis)
d. end-of-life (EOL) products (metal content)
e. EOL metal collected for recycling

f. EOL metal collected for 'downcycling'
g. recycled EOL metal (old scrap)
h. scrap from Manufacturing (new scrap)
j. scrap used in Fabrication (new & old)
m. scrap used in Production (new & old)

(a)

■ <1%   ■ 1-10%   ■ >10-25%   ■ >25-50%   ■ >50%

(b)

Figure 7.4. (a) Flows related to the life cycle of metals and the recycling of production scrap and end-of-life products. Boxes indicate the main processes (life stages) (International Resource Panel, 2011). (b) End-of-life functional recycling rates of sixty elements, with the individual elements categorized into one of five ranges (Reck and Graedel, 2012).

rates above 50%. (The estimates are for 2008 and the committee judged most of them to be "not much above 50%".) Perhaps more dramatic are the more than thirty elements with essentially no functional recycling at all. Only a few elements were assigned values between the two extremes. Thus, a majority of the elements employed in technology are used once and then lost to technology forever, a sad fate given the energy and effort expended to acquire them in the first place.

Further Reading

Ciacci, L., B.K. Reck, N.T. Nassar, and T.E. Graedel, Lost by design, *Environmental Science & Technology*, 49, 9443–9451, 2015.

Dahmus, J.B. and T.G. Gutowski, *Environmental Science & Technology*, 41, 7543–7550, 2007.

Gaustad, G., E. Olivetti, and R. Kirchain, Improving aluminum recycling: A survey of sorting and impurity removal technologies, *Resources, Conservation, and Recycling*, 58, 79–87, 2012.

Graedel, T.E., E.M. Harper, N.T. Nassar, and B.K. Reck, On the materials basis of modern society, *Proceedings of the National Academy of Sciences of the U.S.*, 112, 6295–6300, 2015.

Helbig, C., A. Thorenz, and A. Tuma, Quantitative assessment of dissipation losses of 18 metals, *Resources, Conservation, and Recycling*, 153, 104537, 2020.

Inghels, D. and N.D. Bahlmann, Hibernation of mobile phones in the Netherlands: The role of brands, perceived value, and incentive structures, *Resources, Conservation, and Recycling*, 164, 105178, 2021.

International Resource Panel, *Recycling Rates of Metals*, United Nations Environment Programme, Nairobi, Kenya, 2011.

Kral, U., K. Kellner, and P.H. Brunner, Sustainable resource use requires "clean cycles" and safe "final sinks", *Science of the Total Environment*, 461–462, 819–822, 2013.

Krook, J., A. Carlsson, M. Eklund, P. Frändegård, and N. Svensson, Urban mining: Hibernating copper stocks in local power grids, *Journal of Cleaner Production*, 19, 1152–1156, 2011.

Powell, J.T. and M.R. Chertow, Quantity, components, and value of waste materials landfilled in the United States, *Journal of Industrial Ecology*, 23, 466–479, 2018.

Reck, B.K. and T.E. Graedel, Challenges in metal recycling, *Science*, 337, 690–695, 2012.

Reuter, M.A., A. van Schaik, J. Gutzmer, N. Bartie, and A. Abadías-Llanus, Challenges of the circular economy: A material, metallurgical, and product design perspective, *Annual Reviews of Materials Science*, *49*, 253–274, 2019.

Chapter 8

Material Efficiency

Chapter Concepts

- Material efficiency is a set of strategies for reducing the materials needed to supply a given service.
- Product designers, producers, and users each have opportunities to improve material efficiency, through treating existing products differently or by innovating new product designs with reduced mass and environmental impacts.
- In some cases, services that once involved the physical transfer of materials and products can be completely dematerialized and digitized.
- Increasing material efficiency provides a response to many looming constraints in the supply of virgin materials, but requires a new design focus on the choice and utilization of materials, prolonging product lifetimes, disassembly, and remanufacturing.

8.1 Introducing Material Efficiency

For much of human history, physical resources were difficult to obtain, manufacturing was relatively slow and inefficient, and most people had few material possessions. With industrialization, economies of scale, and rising affluence, there has been an explosion in the quantity of materials that each person uses over their lifetime. This rapid increase in material demand has given rise to many of the large scale environmental challenges that we now face: greenhouse gas emissions, land use change, resource scarcity, to name only a few. The production of products is undertaken, however, to address human needs and wants, such as buildings or transportation. An obvious question is "Can humanity's needs and wants still be satisfied but with reduced or minimal impacts?

The industrial ecology response to this challenge is the discipline of *material efficiency*, which can be defined as the provision of services (shelter, mobility, health, sanitation, etc.) with reduced use of materials and decreased environmental impact. In short, doing more with less (materials).

The topic of material efficiency was inspired by the realization that the growing global population and the desire of all people for more rewarding and comfortable lives will present rapidly increasing demand for materials, and thus enhanced environmental impacts. Returning to the IPAT equation from Chapter 1, material efficiency (and energy efficiency) strategies are all about lowering the Technology term by finding ways to reduce the quantity of resources (and their associated emissions) needed to fulfill demand for services.

8.2 Material Efficiency Strategies

Designers, producers, and users can employ a wide range of material and energy efficiency strategies, as shown in Figure 8.1 (adapted from Allwood and colleagues; see Further Reading). Without changing the basic physical form of a product, designers can prolong the life of products and improve material recovery by making them easier to repair or remanufacture. But other strategies are unlocked when designers innovate around new product forms that can provide the same (or enhanced) services. For example, lightweighting of a vehicle can be achieved by substituting high-strength and light materials (e.g., carbon fiber composites or high-strength aluminum alloys) for heavier iron alloys or by changing the shape of the structure while maintaining its strength (e.g., through castellated beams or structural optimization). Lightweighting is used extensively in mechanical and civil engineering where the cost of materials is significant or there is a performance benefit to reducing weight, as in the case of airplanes.

	Same Product	Redesigned Product

Users
Reduce Use
Use More Intensely
Use for Longer
Product Sharing
Resell / Repair

Modularity / Upgrade
Remanufacture

Producers
Property Improvement
Yield Improvement
Remanufactured Components

Precision Manufacturing
Process Integration

Designers
Design for Remanufacture
Design for Durability

Dematerialization
Lightweighting
Replacing Products with Services

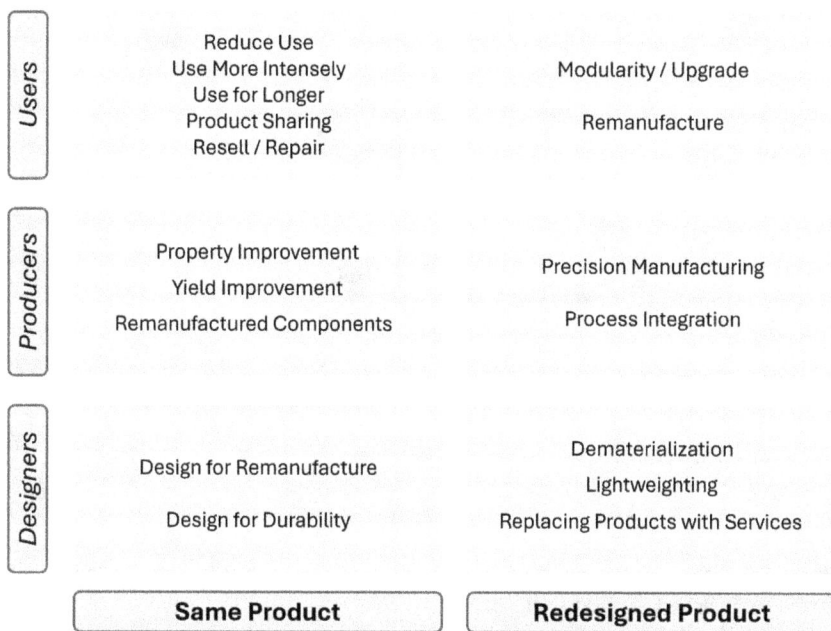

Figure 8.1. Comparing material efficiency and energy efficiency (Allwood *et al.*, 2011).

Designers can also shrink or eliminate components that are redundant or unnecessary, or even invent ways to deliver services that require less or even no materials at all through dematerialization. The history of recorded music is a prime example of dematerialization: songs were first recorded on large wax cylinders, then on large shellac or plastic records, then on smaller magnetic tape spools and cassettes, then on digital compact discs and minidiscs, then digital media players like Apple's iPod, and now largely through streaming services in the cloud, each medium smaller, less material-intensive, and more information-dense than the previous.

Producers also play an important role in material efficiency by upgrading their processes to increase yields (thereby reducing material waste), improve properties of the material feedstocks, or incorporate re-used components. These strategies typically save money by reducing material purchasing costs as well as fees for waste disposal. Finally, users can promote material efficiency by prolonging product lifetimes through reuse and repair or by buying upgradeable products in which only part of the original design is replaced while the rest of the components continue in use.

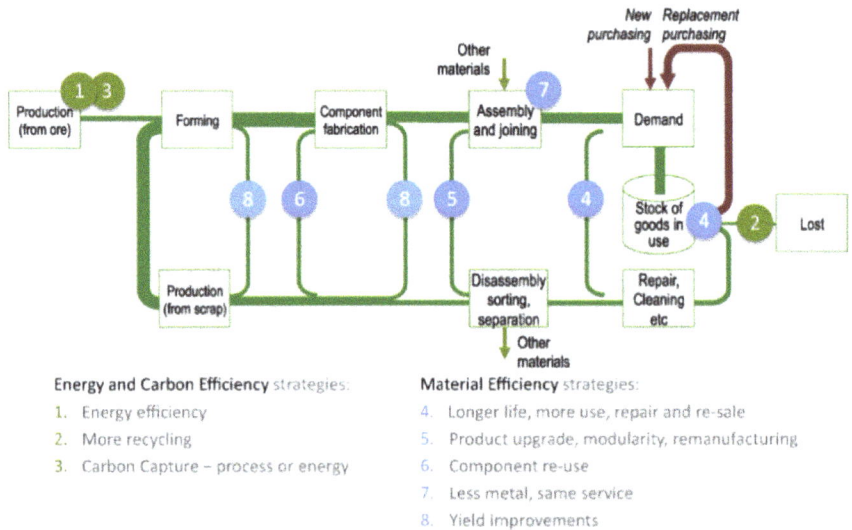

Figure 8.2.   System diagram for the comparison of energy and carbon efficiency with material efficiency (Allwood *et al.*, 2011).

It is also useful to map out the stages in product life cycles during which a material efficiency perspective can be positively employed. A detailed flow diagram for a typical material supply chain from product production to eventual loss is shown in Figure 8.2, in which the numbers in blue represent opportunities to improve material efficiency. Some of these opportunities, such as direct scrap reuse (strategy 8) can occur promptly at the manufacturing site. Alternatively, materials reforming following product discard (strategy 4) may take place years later. Every stage, however, offers the opportunity to reduce original materials input and avoid systemic resource loss. The long-term goal of such an effort, if the goal can be achieved, is the realization of a circular economy (see Chapter 18).

Substitution of one material with another that is less impactful must also account for function, and the properties of materials are, of course, quite different. Those differences demonstrate themselves in the widely known and appreciated "Ashby diagrams", named for materials scientist Michael Ashby of Cambridge University. An example is Figure 8.3, a yield strength *vs.* embodied energy Ashby diagram, which pictures the regions in which different classes of materials fall. Once the technical strength requirements for a material have been calculated by the design engineer, such as

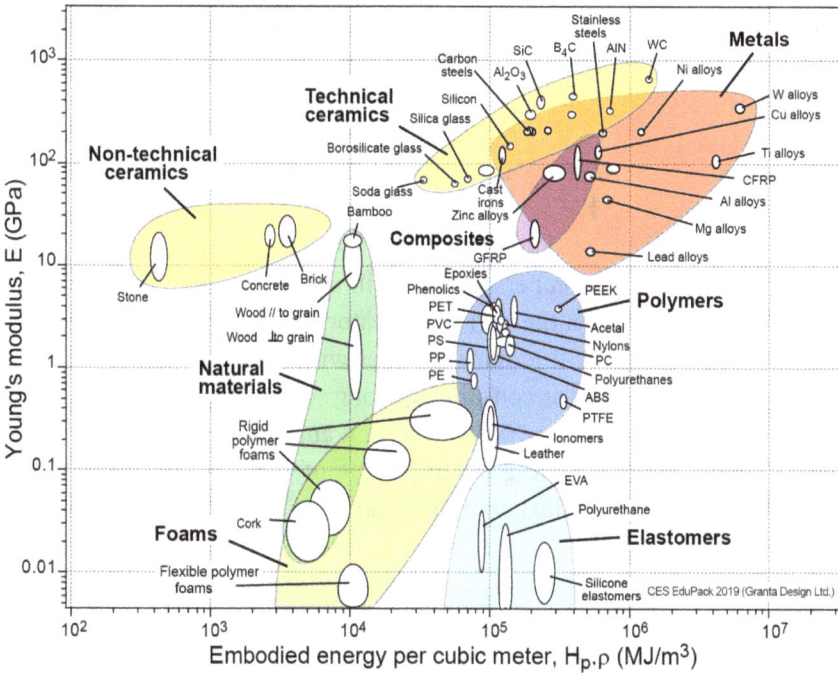

Figure 8.3.   Yield strength as a function of embodied energy, Chart created using CES EduPack 2019, ANSYS Granta © 2020 Granta Design, used with permission.

10^2 MPa on the *y*-axis, then the product design team can consider options starting with the lowest embodied energy on the *x*-axis: first technical ceramics, then composite materials, then metals and alloys. Ashby diagrams can be generated for a large number of physical, chemical, and environmental attributes, and a good product designer will use them and other tools to serve environmental properties of a manufacturing activity as well as to realize a desirable product at its first use.

8.3 Material Efficiency at Scale

A systemic view of material efficiency has been provided by the International Energy Agency (see Further Reading). In that report, material efficiency moves from individual product design to the perspective of the entire economy, recognizing that designing products for long life, prioritizing lightweighting, and promoting increased service life of major construction projects such as buildings and infrastructure can generate both material and environmental benefits. Enabled by public construction standards, practices that focus on reduced material demand, effective maintenance, design for disassembly and reuse, and even compact urban design to minimize sprawl can reduce the material profligacy of modern cities and societies.

This vision of a world designed from both resource sustainability and climate change mitigation perspectives will not be easy to realize because it requires rethinking on the part of educators, corporations, and governments. Such a vision will require political will of governments, personal behavior modifications for consumers, along with a major focus on material efficiency for producers. Nonetheless, it appears inevitable that material efficiency will become a central focus of industrial ecologists and urban planners over the next several decades.

Further Reading

Allwood, J.M., M.F. Ashby, T.G. Gutowski, and E. Worrell, Material efficiency: A white paper, *Resources, Conservation, and Recycling*, *55*, 362–381, 2011.

Ashby, M.F., *Materials and the Environment: Eco-informed Material Choice*. Elsevier, 2012.

Bribeán, I.Z., A.V. Capilla, and A.A. Usón, Life cycle assessment of building materials: Comparative analysis of energy and environmental impacts and evaluation of the eco-efficiency improvement potential, *Building and Environment*, *46*, 1133–1140, 2011.

Cooper, D.R., N.A. Ryan, K. Syndergaard, and Y. Zhu, The potential for material circularity and independence in the U.S. steel sector, *Journal of Industrial Ecology*, *24*, 748–762, 2020.

Cullen, J.M., and D.R. Cooper, material flows and efficiency, *Annual Review of Materials Research*, *52*, 525–559, 2022.

International Energy Agency, *Material Efficiency in Clean Energy Transitions*, Paris, France, 162 p., 2019.

Lifset, R., and M.J. Eckelman, Material efficiency in a multi-material world, *Philosophical Transactions of the Royal Society A*, *371*, 20120002, 2013.

Shanmugam, K. *et al.*, Advance high-strength steel and carbon fiber reinforced polymer composite body in white for passenger cars, *ACS Sustainable Chemistry & Engineering*, *7*(5), 4951–4963, 2019.

Chapter 9

Environmentally-Extended Input–Output Analysis

<table>
<tr><td>

Chapter Concepts

- Input–output analysis uses a sector-based model of an economy to model the links between production and consumption and to analyze how demand for goods and services causes activity throughout the economy.
- Environmentally extended input–output (EEIO) analysis incorporates sector-specific information on environmental burdens to link consumption with measures of emissions, resource use, and environmental impacts.
- EEIO modeling has become one of the most versatile and popular tools within industrial ecology, spawning numerous innovations, including the use of physical input–output tables (PIOTs) to understand material flows through the economy

</td></tr>
</table>

9.1 Understanding Flows at an Economy Scale

Technological societies as a whole are comprised of many factories, not one, and also shops of various kinds, medical facilities, service industries, residences, etc. Complete materials and energy flow data for such systems are not routinely available, but they are essential to a detailed understanding of their resource demands and environmental impacts. On the other hand, transactions between the various entities of society that involve monetary exchanges generate data that is captured by governments.

In the 1930s, the economist Wassily Leontief developed economic input–output (I–O) analysis using industrial sector transaction tables to track these interactions. His approach has proven enormously useful in the field of economics (for which he was awarded the Nobel prize), and it is currently the practice in many countries to routinely generate I–O tables of their national economies. Beginning in the 1970s, economists (including Leontief) and industrial ecologists began to develop methods for including environmental information into I–O tables, so that they could be used to understand how economic activity influences emissions, use of natural resources, and other environmental concerns. This work gave rise to the field of environmentally extended input–output (EEIO) analysis, one of the primary tools of industrial ecology.

9.2 The Mathematics of Input–Output Analysis

Here we will demonstrate the structure of input-output models with a simple two-sector economy, shown in Figure 9.1(a). The variables are:

z_{ij} = monetary flows from sector i to sector j

y_i = monetary flows from sector i entering use (final demand)

x_i = total production (sum of outputs of sector i)

In this arrangement, the elements of each column (z_{1i} and z_{2i}) represent the inputs to sector i, while the elements of each row (z_{i1}, z_{i2}, and y_i) represent the outputs of sector i, and sum to x_i. In traditional I–O analysis, all values are expressed in monetary units, and the number of sectors (equal to the number of rows and columns of the I–O table) may be as many as several hundred, depending on the country.

Import and export flows (also expressed in monetary units) can be included by adding an additional row and an additional column, as shown in Figure 9.1(b). (In this simplified example, we are ignoring other typical I–O model elements such as intermediate inputs and outputs and value added.)

To From	Sector 1	Sector 2	Final Demand	Total Output
Sector 1	z_{11}	z_{12}	y_1	x_1
Sector 2	z_{21}	z_{22}	y_2	x_2

(a)

To From	Sector 1	Sector 2	Exports	Final Demand	Total Output
Sector 1	z_{11}	z_{12}	e_1	y_1	x_1
Sector 2	z_{21}	z_{22}	e_2	y_2	x_2
Imports	i_1	i_2			

(b)

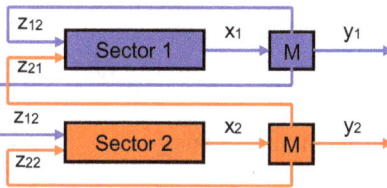

(c)

To From	Sector 1	Sector 2	Final Demand	Total Output
Sector 1	z_{11}	z_{12}	y_1	x_1
Sector 2	z_{21}	z_{22}	y_2	x_2
Env. Flow	F_1	F_2		

(d)

Figure 9.1. (a) A two-sector input–output transaction table. (b) A two-sector input–output transaction table including imports and exports. (c) A representation of an input–output table as a material flow diagram. (d) A two-sector input-output transaction table with appended environmental flows.

Each of the z_{ij} elements represent an input from sector i into sector j that it needs to produce its total output x_j, so we can express the contribution of this input to total output as a proportion:

$$z_{ij} = a_{ij}x_j \quad \text{and} \quad a_{ij} = \frac{z_{ij}}{x_j} \tag{9.1}$$

where the a_{ij} terms are called *input coefficients* or *technical coefficients* representing these proportions.

For the two-sector model, the schematic shown in figure 9.1(a) can be written as a system of linear equations:

$$\begin{aligned} a_{11}x_1 + a_{12}x_2 + y_1 &= x_1 \\ a_{21}x_1 + a_{22}x_2 + y_2 &= x_2 \end{aligned} \tag{9.2}$$

or, expressed in matrix notation:

$$A\bar{x} + \bar{y} = \bar{x} \tag{9.3}$$

where A is a matrix of the technical coefficients a_{ij}, $\bar{y}$ is the final demand vector, and $\bar{x}$ is the production or total output vector. The relationships among sectors, final demand, and total output are shown graphically in Figure 9.1(c). In conventional I–O analysis, A is fixed as it represents the internal structure of the economy.

By definition, the inverse of a square matrix A is a matrix A^{-1}, so that:

$$A \cdot A^{-1} = A^{-1} \cdot A = I \tag{9.4}$$

where I is the identity matrix, and I_2 is

$$I_2 = \begin{bmatrix} 1 & 0 \\ 0 & 1 \end{bmatrix} \tag{9.5}$$

To solve Equation (9.3), subtract $A\bar{x}$ from both sides:

$$\bar{x} - A\bar{x} = \bar{y} \tag{9.6}$$

Grouping terms gives:

$$(I - A)\bar{x} = \bar{y} \tag{9.7}$$

Multiply both sides by $(\mathbf{I} - \mathbf{A})^{-1}$ gives

$$(\mathbf{I} - \mathbf{A})^{-1}(\mathbf{I} - \mathbf{A})\bar{x} = (\mathbf{I} - \mathbf{A})^{-1}\bar{y} \qquad (9.8)$$

Simplifying, this becomes:

$$\bar{x} = (\mathbf{I} - \mathbf{A})^{-1}\bar{y} \qquad (9.9)$$

where the term $(\mathbf{I} - \mathbf{A})^{-1}$ is called the *Leontief inverse* and $\bar{x}$ is the activity vector that describes the total output of each sector of the economy in order to fulfil the final demand $\bar{y}$, including all of the flows z_{ij} among sectors. We can also use Equation (9.9) to calculate how the economy responds when final demand shifts to a new $\bar{y}^*$ by solving for a new total output vector $\bar{x}^*$.

Example 9.1

Consider a simple two-sector economy, where the sectors are Agriculture and Food Processing and the currency is "units". In this economy, the Agriculture sector outputs 10 units to itself (e.g., seeds for planting), 100 units to Food Processing, and 40 units to final demand (e.g., raw food that can be eaten directly). The Food Processing sector outputs 20 units to Agriculture (e.g., to feed farmers), 10 units to itself (e.g., to feed factory workers), and 170 units to final demand. **(a)** Construct an I–O table for this economy.

Now suppose that the population increases and final demand for processed food increases by 50%. **(b)** Use your I–O table from Part **a** to calculate how the total output of each sector will change in response to the increase in demand.

Answer:

Part (a). Following the structure of Figure 9.1(a), we just need to assign which flows are inputs and which flows are outputs. Then, the total outputs from each sector simply sum across each row:

To From	Agriculture	Food Processing	Final Demand	Total Output
Agriculture	10	100	40	150
Food Processing	20	10	170	200

Part (b). The first step is to calculate the technical coefficients a_{ij} using Equation (9.1). a_{11} is the proportion that the flow from Agriculture to itself (10 units) makes up of the total output from Agriculture (150 units), or $10/150 = 0.07$. a_{12} is the proportion that the flow from Agriculture to Food Processing (100 units) makes up of the total output from Food Processing (200 units), or $100/200 = 0.50$. Filling out the remainder of **A** gives:

$$\mathbf{A} = \begin{bmatrix} a_{11} & a_{12} \\ a_{21} & a_{22} \end{bmatrix} = \begin{bmatrix} \dfrac{10}{150} & \dfrac{100}{200} \\ \dfrac{20}{150} & \dfrac{10}{200} \end{bmatrix} = \begin{bmatrix} 0.07 & 0.50 \\ 0.13 & 0.05 \end{bmatrix}$$

Then we can construct the Leontief Inverse in two steps:

$$\mathbf{I} - \mathbf{A} = \begin{bmatrix} 1-0.07 & 0-0.50 \\ 0-0.13 & 1-0.05 \end{bmatrix} = \begin{bmatrix} 0.93 & -0.50 \\ -0.13 & 0.95 \end{bmatrix}$$

$$(\mathbf{I}\quad \mathbf{A})^{-1} = \begin{bmatrix} 1.16 & 0.61 \\ 0.16 & 1.14 \end{bmatrix}$$

(Calculating the inverse of a matrix can be done easily using Microsoft Excel or mathematical software packages.)

Now, let's consider the new final demand:

$$\bar{y}^* = \begin{bmatrix} 40 \\ 340 \end{bmatrix}$$

Using Equation (9.9), the new total output becomes:

$$\bar{x}^* = (\mathbf{I} - \mathbf{A})^{-1} \bar{y}^* = \begin{bmatrix} 1.16 & 0.61 \\ 0.16 & 1.14 \end{bmatrix} \begin{bmatrix} 40 \\ 340 \end{bmatrix} = \begin{bmatrix} 254 \\ 393 \end{bmatrix}$$

Compared to the total output of Part (**a**), the value for Food Processing nearly doubles, but the value for Agriculture also increases by nearly 70%, because even though its own final demand remains constant, it has to supply the Food Processing sector.

Thinking through Example 9.1 raises several practical questions that get to the major assumptions of I–O modeling. First, if production in both sectors increases so dramatically, won't there be economies of scale? In fact, no; I–O models are linear, so that the same quantity of inputs is needed to produce a unit of output, regardless of the scale of production. This is called constant returns to scale. Further, the I–O models assume that the "recipes" for what each sector needs as inputs (represented by **A**) remain constant, called a fixed structure of inputs. Second, won't some food items be affected more than others, depending on the processing required? Because I–O models are aggregate representations of the economy, they are only resolved to the sector level, and can't tell the difference between individual items within a sector. This assumption is called industry homogeneity — each sector has only one recipe, and its output has just one price. In reality, of course processing fruit juice is physically very different than processing chickens, with different amounts of energy and other inputs required, and different prices of the final products. Input-output modeling isn't useful for differentiating between items within the same sector, but it is a fast and convenient method for analyzing economy-wide consequences of consumption.

9.3 Adding Environmental Extensions

The I–O tables of Section 9.2 can tell us much about how the economy responds in monetary terms, but in industrial ecology we want to model environmental flows. This is done by adding environmental extensions or 'satellite tables' to our I–O table that contain flows of interest. For example, we might want to include information on emissions of pollutants, generation of solid waste, inputs of raw materials, or any number of social or biophysical flows. How much are these flows affected per unit of consumption?

Figure 9.1(d) shows an extension of our two-sector economy, now with information on environmental flows F_1 and F_2, representing the totals for that sector. Any number of rows can be added, corresponding to each environmental flow of interest. In the same way that we normalized monetary flows z_{ij}, we normalize the environmental flows F_i by the total monetary output in each respective sector i:

$$f_i = \frac{F_i}{x_i} \qquad (9.10)$$

Then the environmental burdens (scalar) associated with a new final demand vector $\vec{y}^*$ can be calculated as

$$E = \vec{f}\,\vec{x}^* = \vec{f}\,(\mathbf{I} - \mathbf{A})^{-1}\,\vec{y}^* \qquad (9.11)$$

Example 9.2

Let's return to our simple two-sector economy in Example 9.1. Suppose that the Agriculture sector uses 20 TJ of energy per year, while the Food Processing sector uses 1000 TJ. (a) What are the energy requirements per unit of output? (b) If we again double final demand for processed food, how much additional energy will be required?

Answer:

Part (a). Following equation (10.10) the energy required per unit of output is:

$$\vec{f} = \begin{bmatrix} f_1 & f_2 \end{bmatrix} = \begin{bmatrix} \dfrac{F_1}{x_1} & \dfrac{F_2}{x_2} \end{bmatrix} = \begin{bmatrix} \dfrac{20}{150} & \dfrac{1000}{200} \end{bmatrix} = \begin{bmatrix} 0.13 & 5 \end{bmatrix}$$

Part (b). Using equation (10.11), the new energy required is:

$$E = \vec{f}\,\vec{x}^* = \begin{bmatrix} 0.13 & 5 \end{bmatrix}\begin{bmatrix} 254 \\ 393 \end{bmatrix} = (0.13 \times 254) + (5 \times 393) = 1971 \text{ TJ}$$

The total energy use in the base case was 1020 TJ, so this is a near doubling of the energy required to run the economy.

Oftentimes, we want to find the environmental burdens of a single purchase. For this, it is useful to have information on the total environmental flows per unit of expenditure in each sector, denoted $\hat{e}_i$, calculated by setting the final demand $\hat{y}_i$ to just one unit of expenditure in sector i:

$$\hat{e}_i = \bar{f}(\mathbf{I} - \mathbf{A})^{-1} \hat{y}_i \qquad (9.12)$$

Or, calculated as a vector of environmental burdens per unit expenditures $\bar{e}$:

$$\hat{e}_i = \bar{f}(\mathbf{I} - \mathbf{A})^{-1} \mathbf{I} = \bar{f}(\mathbf{I} - \mathbf{A})^{-1} \qquad (9.13)$$

Example 9.3

Again building on Example 9.1, calculate the vector $\hat{e}$ of total energy use per unit of expenditure.

Answer:
We have already done this within our earlier matrix calculations, but here it is explicitly:

$$\hat{e} = \bar{f}(\mathbf{I} - \mathbf{A})^{-1} = \begin{bmatrix} 0.13 & 5 \end{bmatrix} \begin{bmatrix} 1.16 & 0.61 \\ 0.16 & 1.14 \end{bmatrix} = \begin{bmatrix} 0.83 & 5.70 \end{bmatrix}$$

So, if we want to produce 5 units of processed food, the energy required will be 5 units × 5.70 TJ/unit = 28.5 TJ.

Unless you become an industrial ecology researcher, you may make these kinds of EEIO calculations directly, but it is important to understand how they are calculated. It is very common to see EEIO-derived emissions factors built into various carbon, water, and other environmental and ecological footprinting calculators; any factor that is described per unit of expenditure, such as [kg CO_2e per $] for carbon calculators, comes from EEIO modeling.

9.4 The Status and Utility of EEIO Tables

EEIO analysis has advanced tremendously over the past decade and is now being used to quantify economy-level activities across a wide variety of social and environmental issues. Environmental extensions of I–O tables have been accomplished for many countries, regions, and the planet. These EEIO models underpin macro-level calculations of emissions, resource use, and environmental impacts and are the basis for all manner of environmental and resource footprinting efforts. For example, physical input–output tables (PIOTs) can be used with MFA to understand how resources are moving through the economy, and thus how changes in demand or the material content of certain items will percolate through the entire supply chain. The cobalt intensity diagram of Figure 9.2, from Nuss *et al.* (see Further Reading) shows cobalt intensities (in metric tons) for flows between sectors of the US economy. Cobalt is considered by many to be a critical material (see Chapter 16), so the information in the figure helps companies in those sectors understand how their supply chain is dependent on stable, affordable cobalt supplies. Other notable examples of EEIO modeling has revealed the levels of carbon embodied in trade among nations, flows of virtual water associated with the global food system, and in general the relationship between

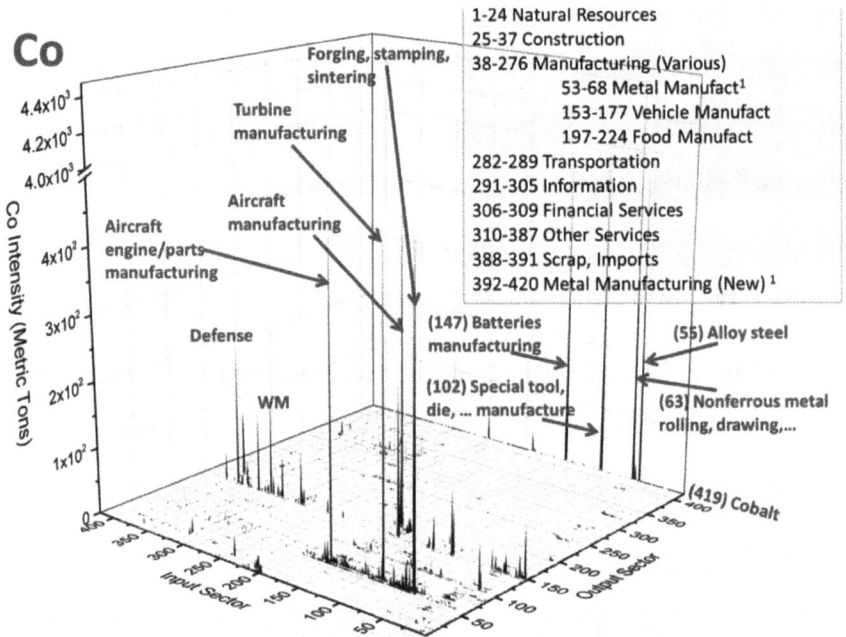

Figure 9.2. A three-dimensional visualization of the physical input-output table of cobalt in the United States, 2007. The horizontal axes are the input sectors (the "sending" sectors) and the output sectors (the "receiving" sectors) of the US economy. The vertical axis is the sector-to-sector cobalt flow intensity in mass units (Nuss *et al.*, 2019).

consumption in one country and production (and its attendant environmental and ecological burdens) in the countries that supply it with various goods and services. Its relevance for quantifying various global issues and the rich data sets that support it makes EEIO modeling is one of the most powerful and flexible tools in an industrial ecologist's toolbox.

Further Reading

Chen, W.-Q., T.E. Graedel, P. Nuss, and H. Ohno, Building the material flow network of aluminum in the 2007 U.S. economy, *Environmental Science & Technology*, *50*, 3905–3912, 2016.

Kitses, J., An introduction to environmentally-extended input-output analysis, *Resources*, *2*, 489–503, 2013.

Lenzen, M., K. Kanemoto, D. Moran, and A. Geschke, Mapping the structure of the world economy, *Environmental Science & Technology*, *46*, 8374–8381, 2012.

Nuss, P., H. Ohno, W.-Q, Chen, and T.E. Graedel, Comparative network analysis of metals use in the United States economy, *Resources, Conservation and Recycling*, *145*, 448–456, 2019.

Charpentier Poncelet, A. *et al.*, Losses and lifetimes of metals in the economy. *Nature Sustainability*, 1–10, 2022.

Shafie, F.A., D. Omar, S. Karupannan, and X. Gabarrell, Urban metabolism using economic input output analysis for the city of Barcelona, in *The Sustainable City VIII*, Vol. 1, *WIT Transactions on Ecology and The Environment*, Vol. 179, 2013.

Tukker, A., G. Huppes, L. van Oers, and R. Heijungs, Environmentally Extended Input-Output Table for Europe, 2006.

Chapter 10

Introduction to Life Cycle Assessment

Chapter Concepts

- Products move through a life cycle, including extraction of raw materials, manufacturing and assembly, transportation, use, and waste management.
- Product life cycles are interconnected through various materials and processes.
- Emissions and environmental impacts can occur at any point in the life cycle of a product; it is often difficult to predict the largest contributors.
- Environmental life cycle assessment (LCA) provides a formalized, quantitative approach to analyzing the environmental impacts of a product or process.

10.1 The Product Life Cycle Concept

The environmentally-related activities of the 1970s and 1980s focused on the manufacturing facility: reducing energy use, waste generation, and various emissions. This was a necessary concentration of effort to address the poor environmental performance of manufacturing at the time, but it considered industrial facilities as isolated entities, completely separate from their suppliers and customers. As we have learned in prior chapters, industrial ecology applies a systems perspective, wherein any industrial activity lives within an industrial ecosystem or network and is highly interconnected to other nodes (facilities, sectors, or users). Any action taken in one node of the industrial ecosystem will influence other nodes, and addressing environmental challenges is best done by considering the overall system.

Focusing greening efforts just at the level of a manufacturing facility ignores critical links in the industrial ecosystem. Any facility needs material and energy inputs for its operations, and often times these resources are extracted and processed elsewhere. The manufactured products may themselves have environmental or human health impacts when used, as well as when they are eventually discarded. In the 1990s, the scope of interest for greening manufacturing was enlarged and the concept of a product *life cycle* began to take hold. Again making use of the biological analogy, physical products began to be considered from "cradle to grave", with associated use of energy, resources, and generation of emissions occurring at every point along the life cycle. Just as we can follow an organism on its physical journey through the world, watching as it interacts with its environment and other organisms, the concept of the product life cycle invites us to follow a product through its own physical journey, watching as its constituent material parts come together from all corners of the globe, as the product is transported, sold, used, discarded, and lives out the rest of its days in a landfill or perhaps transformed by fire in a waste incinerator.

The components of a product life cycle can be defined in various ways depending on the goals and level of detail desired, but the four numbered stages in Figure 10.1 are typical: (1) acquisition/extraction and processing of the necessary resources, (2) manufacture and assembly, (3) use, and (4) end-of-life (EoL), with transport occurring between each of the stages. From any particular point in the product life cycle, activities that occur in previous stages are considered *upstream*, while activities that occur in later stages are considered *downstream*. From the use stage perspective (where most of us interact with products as consumers), upstream corresponds to the production and supply chain of a product, while downstream corresponds to end-of-life.

There are also several opportunities for the cycling of flows among life cycle stages. The generation of reusable discards in manufacturing stimulates a flow of prompt scrap or pre-consumer recycling back to

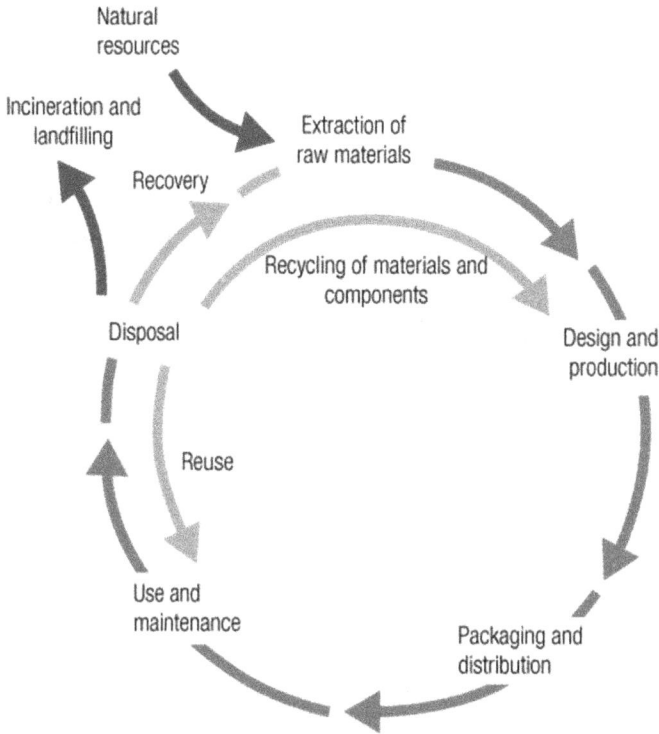

Figure 10.1. Life cycle stages of a generic product (UNEP-SETAC Life Cycle Initiative).

material processing (upper right on the diagram), such as paper mill scrap being reprocessed into pulp. After use (post-consumer), products may still have functionality, stimulating a reuse flow, either for the same purpose or another, such as used cardboard packing boxes being shared or sold. Finally, if the product can no longer be reused directly but the materials still have value for manufacturers, there is a recycling flow of post-consumer scrap for reprocessing into new products.

Consider, for example, the life cycle of paper. Trees are cultivated, harvested, debarked and chipped, and processed into pulp. Next, pulp is used in a paper mill to produce all manner of paper products. That paper is then transported and purchased by consumers for a variety of uses. Some paper may only stay in the use stage for one day before it is promptly discarded, while other paper, such as in a book, will (one hopes) stay in circulation for many years. Finally, paper that reaches the end-of-life phase enters the waste management system, where it could be recycled, incinerated, landfilled, composted, or lost as litter directly to the environment.

We certainly use a lot of paper: an estimated 80–90 million metric tons every year in the United States. How does all of this paper use affect the environment? And more importantly, where should we focus our efforts to reduce these environmental impacts? Traditionally we have focused on manufacturing, for example by improving the energy efficiency of papermaking equipment. By taking a life cycle perspective, we can see many other potential drivers of environmental degradation: deforestation and erosion from overharvesting of trees, water pollution from the pulp plant, particulate matter emissions from the diesel trucks that transport the finished products, or greenhouse gas emissions when waste paper is incinerated.

Which one of these drivers or processes causes the most environmental harm? Should we focus on managing forests better? Should we develop different chemical processes for bleaching paper? Should we invest in new ways to reuse or recycle paper? Or perhaps all the transportation emissions from paper production, delivery, and waste collection is the biggest problem? We want to improve the environmental performance of paper, but how do we know where to focus our efforts?

Actually, this is a very difficult question to answer. First, we need a way to quantify each type of emission so that we can compare them. But there are different types of emissions that cause different types of environmental problems! We can't just compare the mass of a water emission like a bleaching agent to that of an air emission like carbon dioxide, and anyway these emissions are happening in different locations. So, second, we need a way to determine which emissions affect which environmental systems, where, and how much. Finally, third, we need to be able to compare the severity of the impacts on these different systems.

Life cycle assessment (LCA) is a modeling tool that has been developed over the past several decades in order to address the challenges described above. The goal of LCA is to quantify or otherwise characterize all flows of resources and emissions over the entire life cycle of a product, to specify their potential environmental impacts, and to consider alternative designs or approaches that can change those impacts for the better.

10.2 LCA Framework and the ISO Standards

The formal structure and terminology of LCA was promulgated by ISO primarily through standards 14040 and 14044. According to these standards, LCA contains four phases:

(1) *Goal and scope*, where the overall study design is explained and documented;
(2) *Life cycle inventory analysis*, or life cycle inventory (LCI), where flows of resources and emissions are quantified throughout the life cycle;
(3) *Life cycle impact assessment (LCIA)*, where the use of resources and emissions of substances into the environment are linked to physical changes and/or damages; and
(4) *Interpretation*, where results are discussed relative to the original study goal and scope, particularly with respect to issues of data quality and uncertainty.

The four phases of the ISO LCA framework is pictured in Figure 10.2. First, the goal and scope of the LCA are defined and documented. An inventory analysis and an impact assessment are then performed based on this goal and scope. Often the goal and scope must be adjusted because of data gaps or other issues. Finally, the interpretation that follows guides an analysis of potential improvements and uncertainty, which typically feeds back to influence the other three stages, so that the entire process is iterative. It is perfectly normal to redo an LCA several times over the course of a project as adjustments are made to scope, data sources, or other assumptions and methodological choices.

There is perhaps no more critical step in beginning an LCA evaluation than to define as precisely as possible the goal of the study — the precise questions to be answered and for what purpose — followed by choosing the evaluation's scope — what materials, processes, or products are to be considered, and how broadly will alternatives be defined. We will focus on the goal and scope phase in the remainder of this chapter, and will return to the other stages in detail in the following chapter.

Consider, for example, the question of releases of chlorinated solvents during a typical dry cleaning process. The purpose of an LCA on this topic

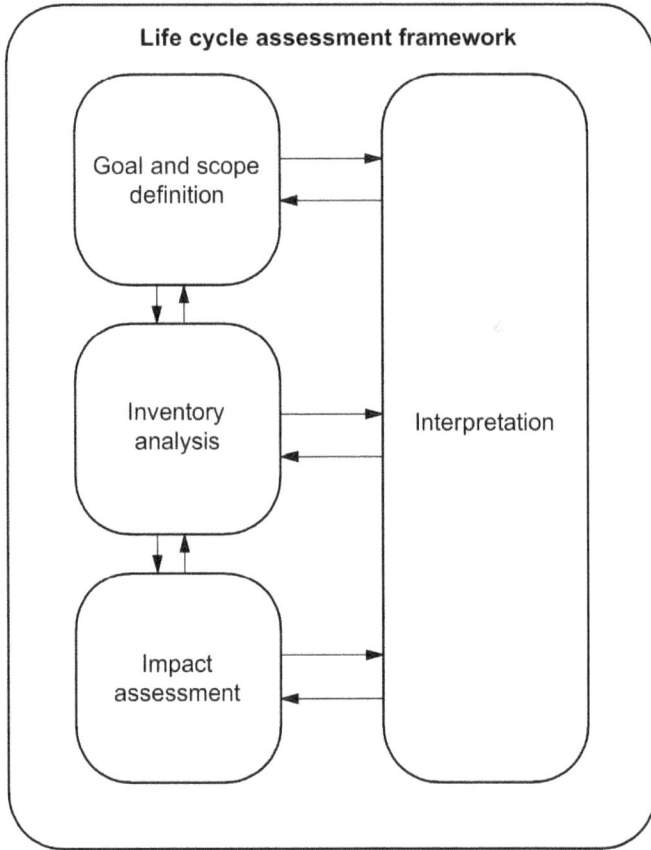

Figure 10.2.   Diagram of LCA phases according to ISO 14040:2006.

might be to identify effective ways to reduce environmental impacts from dry cleaning. But, if the goal and scope of the study are overly narrow, the study might assume that the basic process design is fixed and only consider good housekeeping techniques, end-of-pipe controls, administrative procedures, and perhaps alternative petrochemical solvents. If, however, the goal and scope are defined broadly, the study could include alternative service options. For example, some data indicate that a substantial number of items are sent to dry cleaning establishments not for dry cleaning *per se* but simply for pressing. Accordingly, offering an independent pressing service might reduce emissions considerably. Taking an even broader, systems-level view of the problem, the study might focus on design and consider innovative fiber alternatives or new process technologies that do not require chlorinated solvents at all.

Several methodological and practical considerations influence the goal and scope, including: (a) who is sponsoring and who is performing the analysis, and how much control can they exercise over the implementation of options; (b) what resources are available to conduct the study; and (c) what is the most limited scope of analysis that still provides for adequate consideration of the systems aspects of the problem. The resources that can be applied to the analysis should also be assessed. Most traditional LCA methodologies provide the potential for essentially open-ended data collection — and, therefore, virtually unlimited expenditure of effort. As a general rule, the depth of analysis should be keyed to the degrees of freedom available to make meaningful choices among options, and to the importance of the environmental or technological issues leading to the evaluation.

For example, an analysis of using different plastics in the laptop computer would probably not require complex analysis, because the constraints imposed by the existing design and its market niche make the options that are available to a designer quite limited. On the other hand, a government regulatory agency contemplating rules on single-use plastics used in large amounts in numerous and diverse manufacturing applications would want to conduct a comprehensive analysis, because the degrees of freedom involved in finding substitutes could be quite numerous and the environmental impacts of substitutes could be significant.

10.3 Setting LCA Study Goals

A common LCA objective is to derive information on how to improve environmental performance of a product, or to compare multiple products. If the exercise is conducted early in the design phase, the goal may be to compare two or three alternative designs. If the design is finalized, or the product is in manufacture, or the process is in operation, the goal can probably be no more than to achieve modest changes in environmental attributes at minimal cost and minimal disruption to existing practice. LCA can also be carried out at large scales, comparing not just a single product or process but entire organizations or systems, for example an entire manufacturing facility, corporation, city, or even country. In such cases, the goal might be to evaluate alternative operational designs or to determine what aspects of consumption are causing the largest environmental impacts, in order to inform policy changes. Either way, the objective of a study should match the scale of the problem and the options that the various stakeholders have for implementing change.

According to the ISO standards for LCA, the Goal statement should state and explain the following:

- the aim or objective of the study;
- the intended use of the results, or how they will likely be applied;
- the initiator (or commissioner) of the study;
- the practitioner of the study;
- the stakeholders of the study (interested parties);
- the intended users of the study (target audience); and
- if the results will be used in comparative assertions and released publicly (if yes, this will trigger an obligatory third party peer review).

It is useful to recognize that LCA is an iterative process, and that the goal may need to be revisited as the LCA proceeds.

10.4 Defining the Functional Unit of Analysis

In many cases, we conduct an LCA in order to compare multiple options, so that we might choose the environmentally preferable one. For an eco-conscious consumer, life is full of choices that have environmental implications, and it is often very challenging to know what to do. Is seafood better than red meat? Should I use a sponge or a paper towel or a cloth? Is recycled content paper really better than virgin paper? These types of comparative questions occur for large-scale production systems as well, where the choices affect many people. Should a company build a wood-framed building or a reinforced concrete building? Should the state invest in renewable energy or natural gas-fired power plants? Should society strive to make plastic out of petrochemicals or bio-based chemicals?

In all cases, we need a fair way to compare the options. This is harder than it sounds. Take the sponge vs. paper towel vs. cloth example. Physically these are very different products. It doesn't make sense to compare one of each type, especially because sponges and clothes can be used multiple times and washed, whereas the paper towel is designed for single use. If there is a liquid spill, they each have different levels of absorbance, so you might need one sponge (washed out 10 times) vs. 5 paper towels vs. 2 cloths that then must be washed in the laundry.

LCA recognizes that products have different levels of performance. So, rather than the physical items themselves, it is the core function of the products that is used as the basis of comparison. The quantity of this function you want to compare is called the *functional unit*. As the name implies, the functional unit should specify the common function of the product alternatives, a quantity, and a unit of measure. In the above example, the functional unit could be "the capacity to absorb 1 liter of liquid". The physical quantity of each product alternative that is needed to fulfil the functional unit is called the *reference flow*. This would be the one sponge, five paper towels, or two cloths. In a comparative study, the common functional unit should always be determined first, and the reference flows derived from that functional unit.

The choice of functional unit in large part determines what options can be considered for comparison in an LCA, and so requires consultation and expert judgment. In the example above, if the functional unit is set as 1 liter of absorptive capacity, this excludes options that don't rely on absorption, such as a wet-vacuum or even a thirsty pet! Ideally the functional unit should be set so that the comparison can include all options that could plausibly be taken. The broader the functional unit, the greater the diversity of options that can be considered, including system-level options that may not be immediately apparent.

A functional unit can also be not specific enough. Again using the above example, suppose the liquid that is spilled is a strong acid. Paper towels might disintegrate, and personal safety is also a concern. A more appropriate functional unit would then be "the ability to clean up 1 liter of pH 2 liquid while maintaining personal safety". Time is another aspect of performance that can be specified in a functional unit. One option for dealing with a liquid spill is simply to wait for it to evaporate (assuming the liquid is volatile enough). But this might take hours and not actually be a plausible option. So, in order to exclude this option, a new functional unit could be "the capacity to absorb 1 liter of liquid in under 5 minutes". It is also possible that a product system has more than one function, such as a window that provides a physical barrier but also transmits light. When considering options, both of these functions could be considered in specifying alternative reference flows.

Let's consider another example. An early topic of LCA was the question of shopping bags, a classic question of paper vs. plastic vs. reusable cloth bags. Comparing these bags one-to-one-to-one is incorrect, because each bag provides a different level of performance, a different quantity of stuff that it can hold. Therefore, possible functional units could be "the ability to carry 10 kg of stuff" or "the ability to carry 15 L of stuff". This might require 1 standard #12 kraft paper bag or two typical LDPE plastic grocery bags or a large cloth bag, which would be the reference flows. Since the cloth bag can be reused (this assumes the paper and plastic bags are single-use), its reference flow needs to be adjusted by the number of expected uses. If the average large cloth shopping bag is used 100 times, then its reference flow in this case would be 1/100 bags, with some consideration for cleanliness.

What if we want more breadth? Specifying that the stuff must be carried implies that you have gone to a store to purchase items. If the functional unit instead states "the delivery of 10 kg or 15 liters of stuff to my home", now we can begin to consider larger systems-level questions, such as how to get to the store or e-commerce and drone delivery options. Or, what if we want more specificity? Paper bags perform poorly for wet items, and plastic bags perform poorly for sharp items. If the stuff in question has physical characteristics that preclude certain options, then performance around these characteristics should be specified in the functional unit as well.

Correctly specifying functional units and reference flows is fundamental to any comparative LCA, but can be challenging because it requires us to consider products in a different way than we are used to doing, to see them for what they do rather than what they are. Ideally, the choice of functional unit should be made with input from multiple stakeholders, and revised during the study if necessary. For product systems that have been studied extensively, such as water supply or fuels, consensus functional units have been agreed upon over time.

10.5 Determining an Appropriate Scope of Analysis

As with the Goal statement, the description of the Scope of an LCA must address specific topics according to the ISO standards. These include:

- the functional unit;
- the system boundary;
- major methodological choices, such as mode of analysis and allocation procedure;
- data sources and data quality requirements, including temporal, spatial, and technology coverage;
- impact assessment methods and categories of impacts considered;
- uncertainty analysis;
- other major assumptions.

We will address questions of data and methods in the following chapters on the Inventory, Impact Assessment, and Interpretation phases, and focus the remainder of this chapter on the system boundary.

The potential complexity of comprehensive LCAs is nowhere better illustrated than by the challenge of defining appropriate boundaries of the study. There have been many potential considerations and heuristics developed over the years in order to guide LCA practitioners, but in the end the choice of system boundary should depend on the questions that the study is trying to answer. If the goal of an LCA includes a comparison with other studies, then it is important that study boundaries be consistent. (Indeed, in many sectors there is general agreement about a minimum scope of inclusion for this reason.) But if the initiator of an LCA has specific questions, especially if the results are being used for internal decision-making or research purposes, then the system boundary is more flexible, and is often more targeted. And remember, LCA is an iterative process; your choice is not fixed. Your initial study boundary may need to change as the project evolves, and during the Interpretation phase, it is common to alter boundary assumptions in order to see the effect on overall results.

10.5.1 *Life cycle stages*

The first basic scoping question is which life cycle stages to include. There are three common approaches, and a host of variations; the best choice depends on the goal of the study. Consider again the life cycle diagram in Figure 10.1. Often, we want to know what is the impact of making a product, including raw material extraction, processing, manufacturing, and assembly. An LCA that focuses just on the production stages is called *cradle-to-gate* ("cradle" being natural source of raw materials and "gate" referring to the boundary of the manufacturing facility). This scope can be further truncated to exclude the supply chain and focus only on what happens during manufacturing and assembly stage, which is called *gate-to-gate*. The supply chain is really a sequence of gate-to-gate processes, and these can be strung together to form a cradle-to-gate scope if desired. Finally, if the use stage and end-of-life stage are included, covering the entire product life cycle, this is called *cradle-to-grave*, borrowing a metaphor from the human life cycle. If end-of-life consists mostly of reusing or recycling, then this is sometimes called *cradle-to-cradle*.

A cradle-to-grave study is clearly the most comprehensive, but is often not necessary or even possible. For example, a manufacturer will have access to information and a degree of control over manufacturing and assembly stage and also on the extraction and processes stage (the supply chain) of the various inputs required, but often not, however, on exactly how the product is used or treated at end-of-life. This is especially true if the product is an intermediate such as a polymer resin and is used in a wide variety of products. For these reasons, most manufacturer LCAs are cradle-to-gate. This scope is probably the most common and has different forms depending on the product type. For smaller consumer items, this would include production of the finished good, usually with packaging. For buildings and infrastructure, this would include production of all the materials as well as construction and commissioning.

If the location of use is known, then a cradle-to-gate scope can be extended to include transportation, distribution, and/or sale. For example, for fossil fuels (whose industrial life cycle begins at an oil or gas well), a cradle-to-gate study is called "well-to-gate". If distribution to a fuel station is included, the scope becomes "well-to-pump". If filling up the fuel tank of the vehicle is included, the scope becomes "well-to-tank". For consumer goods, this scope including everything upstream from the point of sale might be called "cradle-to-shelf".

At the other end of the life cycle, LCA comparisons of waste management options usually have the starting point of waste generation, and don't need to consider how various products were used before they entered the waste management system. Or sometimes the use stage is the only stage of interest, for example when different operational modes are being compared and the production and waste management stages are assumed to be identical across options.

The inclusion or exclusion of life cycle stages largely determines what kinds of questions can be answered with an LCA. This can be demonstrated clearly with the example of food. As one of humanity's common basic needs, food and food systems have been analyzed extensively using LCA (and other IE tools). LCA studies including the cultivation and harvesting stage can inform choices around crop types, farming practices, or land use and answer questions such as, is organic farming better than conventional farming? Alternatively, many people are interested in the benefits of local food systems and questions of food-miles and access to healthy food, and to answer these questions, transportation and distribution stages must be included in the study scope. Other researchers focus on new food technologies such as alternative proteins or lab-grown "meat", which are typically analyzed and compared to conventional options using a cradle-to-gate scope. Still other food researchers focus on the topic of food waste, which can occur at multiple points and requires a much broader system boundary, including waste during harvesting (extraction stage), from food processing facilities (manufacturing stage), from spoilage during transportation and storage or from unsold and discarded food, or from waste during preparation or from uneaten food (use stage). Then understanding the environmental consequences of food waste necessitates including the end-of-life stage, depending on if the food waste is composted, landfilled, fed to animals, and so on.

Whatever the topic of an LCA, the choice of which life cycle stages to include should be clearly stated and justified based on the underlying goal of the study and the questions it is trying to answer. Especially for comparative studies, it is critical to include any life cycle stages that differentially influence the various options. Excluding relevant life cycle stages is one of the most common mistakes in LCA. For example, an LCA on sustainable packaging that is only cradle-to-gate and leaves out the end-of-life stage will ignore the fact that packaging materials have very different physical fates, that paper in a landfill will degrade, that paper and plastic in an incinerator will release carbon dioxide (and potentially other air pollutants), that different materials have different recovery and recycling rates; in short, the study will miss important physical flows and have misleading results.

10.5.2 *Physical vs economic modeling of product systems*

So far, we have considered product life cycles in isolation, but in reality they are highly interconnected. A piece of paper is not only made of wood pulp, but also has chemicals added for brightening or coating, whose production also must be considered in the product system. But look beyond just what is physically in the paper — each life cycle stage requires a whole variety of inputs. Just to make the pulp requires processing chemicals, fuel for heating, machinery, electricity to run the machinery, and the physical building that houses all the equipment, which itself is made up of steel, concrete, and other building materials. A pulp plant also requires workers, who require transportation for commuting, food, clothing, protective equipment, on-site bathrooms, telecommunication systems, and computers. The business of the plant also needs insurance, banking services, government permits, advertising, and a host of other services. And in turn, any one of these inputs into the pump plant also requires an entire supply chain of interlinking goods and services. Trace the life cycle for just about any product or service, and the picture quickly becomes quite complex, often requiring us to model processes all throughout the economy. If we want to understand the environmental impacts of a piece of paper (our reference flow in this case), then we need to consider the network of processes that are required to produce it. This is called the *product system*.

There are two basic approaches to modeling a product system in LCA. The first approach is to model the system in physical terms, as a network of physical processes that produce goods and services. This is called *process LCA.* This is typically how engineers are trained to understand the world, as each process can be measured and modeled in terms of physical flows of energy and mass. The second approach is to model the system in monetary terms, as a network of economic sectors that produce goods and services. This is typically how economists are trained to understand the world, as each process can be measured and modeled in terms of monetary flows. We learned the basics of this approach in Chapter 9 on environmentally-extended input-output models. When applied to studying product systems and estimating environmental impacts, this approach is called *economic input–output LCA* (IO-LCA).

Process LCA and IO-LCA each have their own advantages and disadvantages, but one of the most important differences, is how the choice of which to use affects the system boundary of a study.

10.5.3 *Level of detail boundaries*

How much detail should be included in an LCA? An analyst frequently needs to decide whether effort should be expended to characterize the environmental impacts of trace constituents such as minor additives in a plastic formulation or small brass components in a large steel assembly. With some modern technological products containing hundreds of materials and thousands of parts, this is far from a trivial decision. This quandary also applies to considering processes in the life cycle, such as infrastructure. For example, almost all product life cycles include transportation of some form or another, and it is common to include emissions from trucks, trains, planes, and ships, but is it also necessary to consider the environmental impacts of road materials? What about from the production of the machinery needed to build the road? What about the factory needed to build the machinery? What about the transportation of the factory worker? And so on.

In process LCA, where each physical process is modeled separately, it is essentially impossible to include interconnections of every physical input at every level, either because data aren't available or there is simply not enough time. So, a system boundary must be drawn that cuts off some inputs and interconnections; the challenge is to do so in a way that ensures that the results are still accurate, by minimizing what is called *cut-off error*. One rubric for drawing a system boundary is called a *cut-off rule*. If a material or component or process comprises less than a certain percentage of inputs, say 5% by weight or energy or some other physical basis, it can be considered outside the scope of an LCA. The assumption behind the cut-off rule is that weight or energy content is an accurate proxy for environmental impacts. This assumption does not always hold, particularly when there are inputs or components with particularly severe environmental impacts. For example, the lead-acid battery in an automobile weighs less than 5% percent of the vehicle, but the toxicity of lead implies that it should be included if the goal of the study includes toxicity concerns. There are many examples of substances that have a small physical flows but high potency, such as ozone-depleting fire suppressants or radioactive materials, so the LCA practitioner must be very careful not to exclude these important flows from the system boundary as they might actually drive differences among the results. Another type of cut-off rule is to exclude any flow that contributes less than a certain percentage of environmental impacts. To make this determination requires some initial analysis, but then inconsequential but time-consuming flows can be safely left out of the LCA.

Another rubric for setting the system boundary of a process LCA is to leave out certain categories of inputs, with the implicit assumption that they do not contribute much to environmental impacts and their exclusion will not meaningfully affect the results. This exclusion rubric is commonly applied to inputs of labor, machinery, or infrastructure. Note that this is distinct from the exclusion of life cycle stages. Rather, the exclusion of, say, machinery, from an LCA would apply to all life cycle stages, including machinery needed for extraction, production, transport, use, and end-of-life. The risk here is that the excluded category actually is important for the product system being studied. For example, for an LCA that compares different freight transportation options, it would be prudent to include infrastructure (but still perhaps exclude labor). Whatever the system boundary is used, any flows that fall outside of the boundary of a process LCA need to be explicitly documented in the study scope.

While each individual flow that is cut off in a process LCA might be small, their cumulative effect can be significant. In contrast, IO-LCA completely avoids the cut-off problem. Because IO-LCA models the monetary flows and emissions over the entire economy, all of the inter-connections among processes (sectors) are included and nothing is left out. IO-LCA has other disadvantages such as aggregation and a lack of product specificity, but one major advantage is its comprehensiveness.

10.5.4 *Spatial and temporal specification and boundaries*

For every LCA, it is essential to know where and when the product system is being considered. This information is best provided alongside the reference flow, such as "1 kg of tomatoes grown in the Netherlands and shipped to the UK in 2020". The specification of location and time period in the study Scope serves as a benchmark to judge the quality of the data used in the LCI. Tomatoes grown in the Netherlands may require different types and quantities of inputs than those grown in Spain, and of course shipping to the UK is closer than to, say, South Africa, requiring less fuel. And one hopes that both cultivation and shipping have improved in resource efficiency over time. The locations and timing of production and consumption indicates exactly which product system is under study.

In addition, the study scope should specify not just where and when emissions are taking place, but also the boundaries of the environmental impacts that are caused. A characteristic of environmental impacts is that their effects can occur over a very wide range of spatial and temporal scales. The emission of large soot particles mostly affects a local area, those of nitrogen oxides generate acid rain over hundreds of kilometers, and those of carbon dioxide influence the entire planetary energy budget. Similarly, emissions causing photochemical smog have a temporal influence of only a day or two, the disruption of an ecosystem several decades, and the stimulation of global climate change several centuries. Which spatial and temporal boundaries to choose again depends on the goal of the study, but in any case these boundaries should be clearly stated in the Scope of an LCA. If, for example, an LCA is being conducted by a national agency with the goal of assessing domestic health impacts, then a national spatial boundary would be suitable. But if the goal is to understand the environmental impacts of a new technology, then considering effects on global ecosystems would be appropriate. Unsurprisingly, the larger and longer the system boundaries are, the more time-consuming and scientifically uncertain an LCA becomes. Exactly how the spatial and temporal boundaries of environmental impact are modeled is part of the LCIA phase of LCA.

10.5.5 *Choosing appropriate and consistent boundaries*

It should be apparent that the choice of LCA boundaries can have enormous influence on the time scale, cost, results, meaningfulness, and tractability of an LCA. The best guidance that can be given is that the boundaries should be consistent with the stated goals of the study. The most common mistake made in setting the system boundary is simply omitting important information. A vague system boundary makes it difficult or impossible to interpret results properly or compare them to other product systems or LCA studies. Clear documentation is essential for the system boundary and throughout the goal and scope phase of an LCA and is a primary reason for the existence of international standards.

Further Reading

Guinée, J. *et al.*, *Handbook on Life Cycle Assessment — Operational Guide to the ISO Standards*, Dordrecht, The Netherlands: Klewer Academic Publishers, 2002.

Guinée, J. *et al.*, "Life cycle assessment: Past, present, and future". *Environmental Science & Technology, 45.1*, 90-96, 2011.

Matthews, S., C. Hendrickson, and D. Matthews. *LCA Textbook*. Online at: https://www.lcatextbook.com/.

UNEP-SETAC Life Cycle Initiative. Online at: https://www.lifecycleinitiative.org/.

Chapter 11

Life Cycle Assessment in Practice

Chapter Concepts

- The fundamental building block of an life cycle assessment (LCA) is the unit process, each of which has discrete inputs and outputs, and which are linked together to form a product system.
- The life cycle inventory is the complete list of material and energy flows for a product system, and is made up of foreground data (collected directly) and background data (linked through existing datasets).
- Life cycle impact assessment (LCIA) quantifies how flows in the inventory cause environmental impacts, either in terms of environmental changes (midpoint) or damages (endpoint).
- Proper interpretation of the results includes consideration of uncertainty and reflection on whether the LCA has adhered to its goal and scope; iteration is common.

11.1 Building a System Diagram

Once the goal and scope of an LCA have been established, including the definition of functional unit and the reference flows for each of the options being considered, an LCA practitioner can begin the modeling process. The first step is to construct, ideally in cooperation with the design and manufacturing team, a *system diagram* that includes all the life cycle stages included in the study, and to break each stage down into *unit processes* that represent each major step or transformation.

Each unit process within the product system is commonly represented as a block, with arrows representing flows of materials, energy, or emissions (Figure 11.1). There are two types of flows that we want to track. *Economic flows* are those that are linked to other unit processes (that is, they are inputs from or outputs to other parts of the economy). Inputs of electricity and processed materials or outputs of solid waste sent for treatment are all examples of economic flows. In contrast, *environmental flows* are those that are linked directly with nature (that is, they do not connect to other unit processes in the product system). Water taken directly from a river for process cooling or pollutants released directly to the atmosphere are examples of elementary flows. It's not necessary to label flows in this way in a system diagram, but this distinction will become important when we look at the computational structure of life cycle assessment below.

Unit processes are then linked together by various flows; the resulting system diagrams are thus commonly referred to as block diagrams or process flow diagrams (PFDs), and are extensively used in the various fields involving process engineering. System diagrams will also show processes that are excluded from the study, either because they are not relevant to

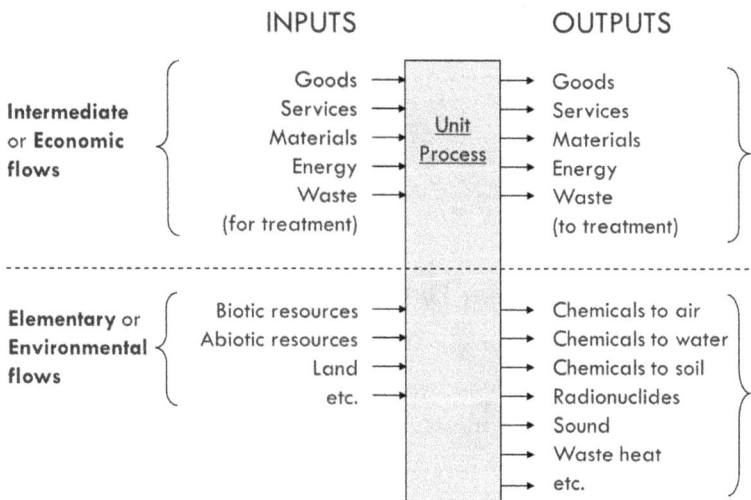

Figure 11.1. A generic diagram of a unit process (adapted from Geyer).

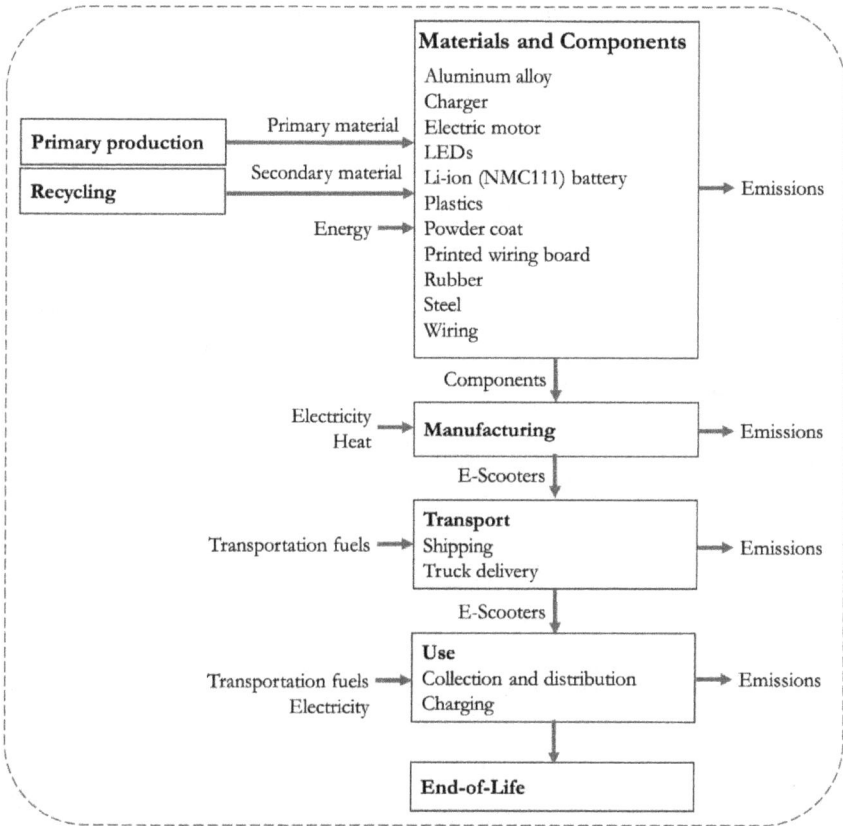

Figure 11.2. A system diagram of the life cycle of an electric scooter (Hollingsworth *et al.*, 2019).

the goal, or for some other methodological or practical reason. Figure 11.2 shows a system diagram for the life cycle (cradle-to-grave) of an electric scooter (see Hollingsworth *et al.*, 2019). We will return to this example throughout the chapter to reflect on each phase of an LCA.

11.2 Compiling the Life Cycle Inventory

11.2.1 *Acquiring foreground data*

Once the system diagram of the LCA has been established, the analyst proceeds to collecting the necessary data on input and output flows, during is called the life cycle inventory phase of LCA (see Figure 10.2). This phase typically begins with examining the main product or process reference flow, either by taking measurements directly or by working with manufacturers or public information. What is the product made of, and what are the quantities of different materials? This information could be found by deconstructing the product and weighing the pieces, or from a manufacturer's specifications sheet. If the product is assembled from components supplied by others, it may be necessary to deal with suppliers to get a complete picture. How much electricity is used to make the product? If the factory is accessible, this value could be measured directly with individual meters, or derived from the factory's electricity bills. How far does the product travel to the user, and how? This value can be estimated from online maps, calculated based on internal details from a manufacturer's logistical operations, or averaged from freight transport data reported to the government. Whether measured directly, calculated based on engineering models, estimated, or averaged, these product- or process-specific inputs and outputs are called *foreground data*, as they are in the foreground of the product system, "within sight" of the LCA practitioner. Foreground data can include both economic and environmental flows.

In the case of the e-scooter study, researchers disassembled and weighed a scooter, estimated shipping distances from the manufacturer, estimated electricity needed for charging based on manufacturer specifications, used data from the e-scooter sharing platform to understand how much the scooters were actually being used, and conducted a survey of employees to understand the extent to which scooters needed to be relocated after hours. This mix of direct measurement, technical specifications, estimation, company data, and personal interviews is typical of the foreground inventory analysis phase of LCA.

11.2.2 *Linking foreground data to background data and LCI databases*

Foreground data represent only *direct* inputs and outputs individual stages of a product life cycle, but they don't yet tell us anything about the product system as a whole, in particular all of the indirect flows that occur upstream in the supply chain or downstream in waste management.

For example, the amount of electricity used annually in a building doesn't reveal how that electricity was generated, how much fuel was used (if non-renewable), what emissions were generated and where, the quantity of water used at the power plant, etc.

The second step of compiling the life cycle inventory is to connect each economic flow in the foreground either upstream or downstream to other unit processes in the product system. It is not practical for an LCA practitioner to investigate every process and supplier along every branch of a supply chain; there are so many interconnections in our global economy that the study would never be completed!

Luckily, there are already inventory data for thousands of unit processes that have been investigated in the past, based on decades of LCA research. So, rather than an LCA practitioner collecting data directly from each individual material supplier, the product-specific economic flows in the foreground can simply be linked to existing LCI data sets, called *background data*. Background LCI data sets live in LCI databases that are compiled by researchers, government agencies, or commercial entities. There are dozens of LCI databases globally; taken together, they represent a giant intellectual contribution to the field. A list of common LCI datasets can be found on the openLCA Nexus page (https://nexus.openlca.org/databases). In our example, the e-scooter researchers linked the quantity of aluminum needed for the scooter with a background dataset name "Aluminum alloy, $AlMg_3$" from the ecoinvent 3 LCI database. All LCA studies should clearly document how foreground data were matched to background data so that these matching choices can be checked and validated.

Once all of the foreground economic flows have been linked to background LCI data sets, an algorithm (described in the next section) traces each supply chain all the way back to the initial extraction of resources from nature and counts all of the emissions, scaling each unit process according to the quantities needed to produce the final reference flow. This system-wide set of resource inputs and emissions is called the life cycle inventory and may cover thousands of individual substances or flows.

The life cycle inventory forms the basis of life cycle impact assessment (LCIA) (described in Section 11.3), but can also provide valuable environmental insights, especially about resource inputs. For instance, the life cycle inventory can tell us how much energy is used throughout the life cycle of a product, both directly and indirectly — this is called *primary energy demand* or *cumulative energy demand*. We can also see how much water is withdrawn from natural sources throughout the life cycle — this is a form of *water footprint*. Total land use, total emissions of a specific pollutant, total solid waste generated, all of these can be extracted from the life cycle inventory and used as a basis to compare product options or to target environmental improvements.

One fascinating inventory analysis study considered the mass of inputs required to manufacture a 2 g microchip (see Williams *et al.*, 2002). The authors found that the chemical input needed to make the chip (including upstream use of fossil fuels) weighed 1.7 kg, or 630 times the mass of the chip itself! Water used for fabrication contributed an additional 32 kg of material use. Chip fabrication relies on high purity materials and highly controlled process environments, and based on these results, we say that the chip has a large material footprint. This type of analysis has been carried out for numerous products, and it is always striking to see that the upstream use of resources is typically many times larger than the resources that physically make it into the final product. Figure 11.3 is an artistic (but dimensionally accurate) visualization of this result for a car.

11.2.3 *Computational structure of life cycle inventory analysis*

Each of the unit processes in an LCI database has input and output flows, and remember that these can be grouped into economic flows (linked to other unit processes) and environmental flows (directly to and from nature). Consider a group of n unit processes, each of which can be expressed as a vector $\overline{p}_i$ that has economic flows a and environmental flows b:

$$\overline{p}_i = \begin{pmatrix} a_1 \\ \vdots \\ a_r \\ b_1 \\ \vdots \\ b_s \end{pmatrix}. \tag{11.1}$$

Figure 11.3. Photographic visualization of the embodied materials in a car, represented next to a normal sized car but ~3× the size and ~25× the mass (from the LCAart project, lcaart.com).

There are r possible economic flows and s possible environmental flows. If we line up all n unit processes side-by-side, we can list all of the flows in the database, separated into two adjoint matrices **A** and **B**, with economic flows on top and environmental flows on the bottom:

$$\begin{pmatrix} a_{11} & a_{1n} \\ & \\ a_{r1} & a_{rn} \\ b_{1n} & b_{1n} \\ & \\ b_{s1} & b_{sn} \end{pmatrix} = \begin{pmatrix} A \\ \overline{B} \end{pmatrix} \tag{11.2}$$

A is called the *technology matrix* and describes how the unit processes are linked together through economic flows. Each unit process has a reference flow that describes its primary economic output and each economic flow a_i has a corresponding unit process that produces it, so that **A** is square and

$r = n$. **B** is called the *environmental matrix* and describes how the unit processes are linked to nature. There may be any number of environmental flows, so there is no requirement that **B** be square.

The basic matrix mathematical problem when calculating the life cycle inventory is to trace how each of the foreground inputs that make up the reference flow (the product or process under study) causes activity in upstream or downstream unit processes. These foreground inputs can be expressed as a final demand vector $\overline{Y}$:

$$\overline{Y} = \begin{pmatrix} y_1 \\ \vdots \\ y_r \end{pmatrix} \tag{11.3}$$

We want to "activate" each unit process in the technology matrix **A** in order to deliver these economic flows. In other words, we want to find the vector of activity coefficients $\overline{X}$ such that:

$$A\overline{X} = \overline{Y} \tag{11.4}$$

Solving for $\overline{X}$ gives:

$$\overline{X} = A^{-1}\overline{Y} \tag{11.5}$$

So, the economic flows of each unit process are scaled by the corresponding activity coefficient in order to deliver the economic flows that constitute the final demand. At the same time, the environmental flows of each unit process are also scaled by the same activity coefficients. We can thus express the system wide environmental flows as

$$\overline{E} = \mathbf{B}\overline{X} = \mathbf{B}A^{-1}\overline{Y} = \begin{pmatrix} b_{11} & \cdots & b_{1n} \\ \vdots & \ddots & \vdots \\ b_{s1} & \cdots & b_{sn} \end{pmatrix} \begin{pmatrix} x1 \\ \vdots \\ xn \end{pmatrix} = \begin{pmatrix} \sum\limits_{i=1}^{n} b_{1i} \cdot x_i \\ \vdots \\ \sum\limits_{i=1}^{n} b_{si} \cdot x_i \end{pmatrix} \tag{11.6}$$

$\overline{E}$ is the life cycle inventory vector, it represents the quantity of each type of environmental flow summed over all of the unit processes in the product system. For example, $\overline{E}$ could contain the total amount of freshwater withdrawn as a resource input or the total amount of carbon dioxide emitted from the product system. For more detail on the computational structure of LCA, see Heijungs and Suh in the Further Reading.

11.2.4 *Allocation and system expansion*

One complication in compiling a life cycle inventory is when a unit process has more than one reference flow, that is, more than one salable product output. Actually, this is quite common in industry; for example, refineries can produce a dozen or more different fractions of chemicals using crude oil as an input. The refinery has resource inputs and emissions, so how best to assign, or *allocate* those flows to the multiple product outputs?

The ISO standards for LCA provide specific guidance on this topic, giving a list of options, detailed below and shown in Figure 11.4.

(1) *Avoid allocation by sub-dividing the unit process*: In some cases, a unit process might have discrete inputs and output for each of its outputs. For example, a food processing plant could grind oats into

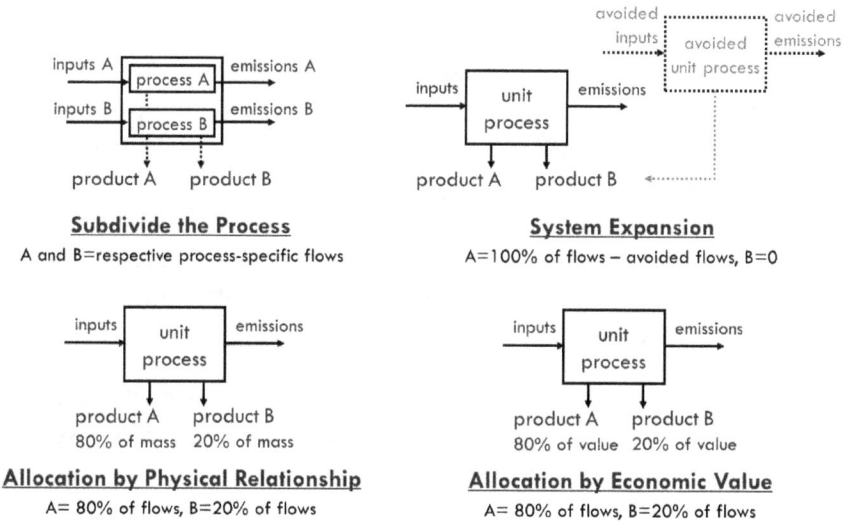

Figure 11.4.   Approaches to allocation.

oatmeal and corn into cornmeal using different machinery. As long as each of the inputs and emissions can be assigned to one and only one of the output products, then we can break the multi-output unit process into multiple single-output unit processes.

(2) *Avoid allocation by system expansion*: This is the most complex option, one that forms an entire sub-discipline within LCA. The idea here is to choose a primary output to which all of the flows are allocated, but then also assign it credits (negative flows) because the other output(s) can act as substitutes and avoid production. In this way, the system is expanded to include the avoided production as a credit. For example, a co-generation plant produces both electricity and steam heat. An alternate way of generating that heat is with steam boilers. Using system expansion, we would assign all of the fuel inputs and emissions to the electricity, but then subtract the avoided emissions from the steam boilers because they no longer need to operate.

(3) *Allocation on the basis of physical relationships*: In this case, we assign all flows of the multi-unit process to each product output according to its proportion of some total physical quantity, typically mass or energy content. So, under mass allocation, if in a two-output process, product A has 95% of the mass, then it is allocated 95% of each flow, and product B is assigned 5%.

(4) *Allocation on the basis of economic value or some other relationship*: The final option is to use a non-physical proxy for allocating flows, the most common choice being economic value. This option resonates with many businesses because industrial facilities are often built with a primary product or source of revenue in mind, while other co-products only marginally influence profitability. The approach is very similar to the earlier allocation example; under economic allocation, if in a two-output process, product A has 95% of the economic value, then it is allocated 95% of each flow, and product B is assigned 5%.

Different industries have developed their own guidelines about which approach to allocation should be followed for LCA studies in their sector. The choice of allocation can greatly affect results, and for this reason, it is critical that an LCA practitioner clearly document how allocation was done.

11.3 Life Cycle Impact Assessment

The life cycle inventory flows tell us only the magnitude of resource inputs and emissions for each option. The third phase of LCA, called LCIA, allows us to quantify effects of these flows on the environment and public health. Whereas the inventory phase is largely about accounting, the assessment phase relies on models from environmental science, chemistry, physics, ecology, toxicology, and other fields in order to create causal chains that link an emission in a particular location to the physical changes (called *midpoints*) or damages (called *endpoints*) that it causes all over the world.

Of course, there are multiple ways that our activities affect the planet: global warming, depletion of the ozone layer, and toxicity are all critical concerns. These are called *impact categories*. One of the key features of LCA is that it considers multiple impact categories simultaneously, in order to provide more complete information for decision-making. This also can avoid *burden shifting*, where we lessen impacts in one category while unintentionally increasing impacts in another. Different models are used to assess different types of impacts; an LCIA method defines which impact categories are included and how each one is assessed.

LCIA methods are typically developed by scientific research organizations within government labs and universities, with an intended focus or geography. Table 11.1 shows an example for the United States, called Tools for Reduction and Assessment of Chemicals and Other Environmental Impacts (TRACI). While other LCIA methods might consider different sets of impact categories and use different models to do so, most methods cover the same themes of resource use, atmospheric change, air quality, water quality, and toxicity.

The LCIA procedure is a four-step process: *classification* and *characterization* are required by the ISO standards, while *normalization* and *weighting* are optional.

Classification: Each flow is classified according to whether it causes a certain type of impact. For example, nitrous oxide (N_2O) causes both global warming and ozone depletion, and so is assigned to both impact categories. Other flows might not cause effects in any of the impact categories, and are left unassigned.

Table 11.1.   Impact categories covered by the TRACI 2.1 LCIA method.

Impact Category	Description	Indicator
Global Warming Potential (GWP)	Indicates the radiative forcing (heat trapping potential) associated with greenhouse gases (GHGs) and other climate forcers, using a 100-yr time horizon	CO_2 eq
Acidification Potential (AP)	Indicates the potential acidification of water and soils due to acid rain, formed by reaction of emissions such as sulfur and nitrogen oxides with water in the atmosphere	SO2 eq
Eutrophication Potential (EP)	Indicates the potential for algae growth and reduced amounts of oxygen in water due to nutrient pollution	N eq
Ozone Depletion Potential (ODP)	Indicates the potential for destruction of ozone in the stratospheric ozone layer	CFC-11 eq
Photochemical Smog Formation	Indicates the potential for formation of tropospheric (surface-level) ozone due to reactions of nitrogen oxides and volatile organic compounds (VOCs) in the presence of sunlight	O_3 eq
Human Health Particulate Matter	Indicates the potential for emissions and secondary formation of harmful particulate matter (PM)	$PM_{2.5}$ eq
Human Health Cancer	Indicates the potential for chemical emissions to cause cancer in humans	CTUh
Human Health Non-Cancer	Indicates the potential for chemical emissions to cause non-cancer disease in humans	CTUh
Ecotoxicity	Indicates the potential for chemical emissions to cause damages to non-human organisms	CTUe
Resource Depletion	Indicates the potential for depletion of non-renewable resources, in particular fossil fuels	MJ eq

Characterization: Each classified flow is assigned a value, or characterized, according to how potent a substance it is for each type of impact. This is called a characterization factor (CF) and is expressed relative to an indicator substance. In the case of global warming potential, we measure CFs relative to carbon dioxide, and so all greenhouse gases are characterized in units of carbon dioxide equivalents (CO_2 eq). Then the mass of each classified flow E_j from the life cycle inventory is multiplied by its respective characterization factor $CF_{i,j}$ and summed up over all j flows to determine the total impact I_i for each impact category:

$$I_i = \Sigma_j CF_{i,j} \cdot E_j \qquad (11.7)$$

Normalization: Because LCA compares options across multiple sustainability indicators, results typically have different units, many of which are unfamiliar to non-scientists. It can be useful to normalize results across impact categories by some set of reference values R_i, for example as a proportion of a person's average annual impacts, in order to tell which type of impact is larger on a relative basis. The normalized results N_i are then:

$$N_i = \frac{I_i}{R_i} \qquad (11.8)$$

Weighting: Valuation is the process of assigning weighting factors to the different impact categories based on their perceived relative importance. For example, an assessor, an international standards organization, or a stakeholder panel might choose to regard climate change impacts as twice as important as acidification, and apply weighting factors w_i to the normalized impacts accordingly. Then the impact assessment results can then be expressed as a single weighted score W:

$$W = \Sigma_i w_i \cdot N_i \qquad (11.9)$$

Single scores are simple and attractive for comparing products and tracking progress. But, the application of weighting factors is controversial because doing so involves making social, political, and ethical choices about the relative importance of different types of impacts. As a result, LCIA evaluations that are suitable for a particular culture or location or time are unlikely to be useful in other circumstances. Because of this limitation, weighting is often omitted from LCIAs. (Doing so, however, is equivalent to making an implicit weighting with all categories valued the same.)

11.4 Interpretation

This last and perhaps most important phase of LCA is interpretation, where results get translated into knowledge and recommendations. If Option A has lower impacts than Option B across all impact categories, then this supports a clear preference. But it frequently happens that one product option is better for some categories, but worse in others. The LCA practitioner must discuss these different trade-offs in the context of the study Goal, leaving the final decisions to the client. Overall, the purpose of the Interpretation phase is to provide context to the study so that readers can fairly interpret the results.

11.4.1 *Draw conclusions and recommendations*

In many cases an LCA is motivated by a desire to improve designs and practices. These studies will include a *contribution analysis* to tell which processes or hot spots are driving results for different impact categories, in order to suggest where to focus efforts. If a particular material is driving impacts, then designers might investigate low-carbon alternatives or ways to reduce how much material is used. If a particular upstream process is the main culprit (such as electricity generation), then managers might work with suppliers to source a greener option (e.g., through renewable energy contracts). This list of recommendations constitutes the core output of an LCA study.

In developing a list of recommendations based on the inventory and impact assessment results, it is important for the assessor to be inclusive, and to range widely. Some recommendations will be very specific (i.e., avoid the use of toxic metals), while others are much more diffuse (i.e., minimize the diversity of packaging materials). Both types are important to include. The highly specific recommendations are easier to generate, and their accomplishment is more easily measured. The diffuse recommendation may be more difficult to deal with, but may in some cases be very important; their inclusion may crucial to actually effecting change.

11.4.2 *Evaluate the data used in the LCA*

LCA is not an exact science — there are various assumptions and choices that are a part of every study, as well as inherent uncertainty in model parameters such as material quantities and emissions factors. An essential step in the LCA interpretation phase is to evaluate the quality, completeness, and consistency of the data. The goal is to ensure that each identification of significance is backed by adequate, reliable information. This is particularly important if alternative product designs are being evaluated, because the designs need to have consistent bases for comparison.

A data quality assessment is an investigation of how well the input data reflect the reality of what is being modeled. For instance, if a study uses emissions factors that are 5 years old but the actual technology has improved dramatically, then this must be clearly documented and its implications discussed. Several dimensions of data quality can be investigated; a typical list of data quality indicators includes:

- *Reliability*: How were the data gathered, from direct measurement, secondary sources, estimation, etc.
- *Completeness*: How well do the data represent the range of production sites for the process or product.
- *Temporal correlation*: How long between when the data were collected and the time of the study.
- *Geographical correlation*: How well does the location of the data source match the location of the study.
- *Technological correlation*: How well do the data represent the actual process or product being modeled.

For each data set used, each of these data quality indicators might be scored 1–5 or described qualitatively to make up the data quality assessment.

At the data completeness step, one wishes to confirm that all product life stages have been addressed, as well as all relevant environmental impacts. This information should be verified as meeting the system boundaries established at the beginning of the study, and that the significant raw materials and releases have been incorporated. Finally, at the data consistency step, the LCA practitioner checks that consistent choices have been made in terms of data quality, source databases and background data sets, allocation, LCIA methods, and other methodological choices across the options being considered.

11.4.3 *Sensitivity and uncertainty analysis*

LCA studies often include a *sensitivity analysis*, where important assumptions are changed in order to see how much the overall results respond. If an LCA is found to be very sensitive to a single parameter, then a caveat should be added to the recommendations and the parameter might be investigated more in order to make sure that the best possible information is being used. In the study on e-scooters, sensitivity analysis was conducted on several parameters, including the daily use and product lifetime. Testing the full range of values for these parameters produced global warming results that spanned a factor of 3. This magnitude of model sensitivity isn't necessarily bad or particularly unusual; it just means that care must be taken in generalizing the results. In addition to testing different parameter values, a sensitivity analysis might also test other choices, such as functional unit, allocation scheme, LCIA method, and so on, in order to make sure that the recommendations coming out of the study are robust, and not unduly influenced by modeling choices.

LCA studies might also include a quantitative *uncertainty analysis* using statistical methods. The most common approach is called Monte Carlo simulation, which aims to combine the uncertainty of each data point into an uncertainty estimate for the overall results. In this approach, each data point is assigned a probability distribution (e.g., a normal distribution) that defines its possible range of values. Then the entire model is run hundreds or thousands of times, with each run sampling a different set of values for each parameter. The distribution of model results can then be used to create confidence intervals, compare options, and run statistical tests. Figure 11.5 shows Monte Carlo simulation results for the e-scooter study under several different modeling scenarios. The mean values of each study are shown by the vertical bars. These values alone *could* be used to communicate results, but also showing the spread of the results adds a new dimension for interpretation and conveys where there is significant overlap among options.

11.5 LCA Software

In this chapter, we have presented the fundamentals of LCA methods and mathematics, but because of the complexity of the LCA process, most practitioners rely on software that holds the background inventory data, allows for easy inputting of foreground data and metadata, and runs the algorithms (including uncertainty analysis). LCA software comes in two

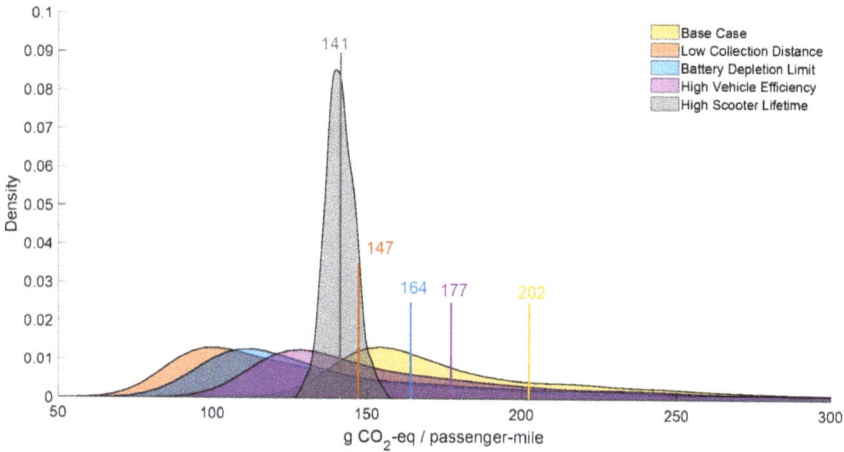

Figure 11.5. Monte Carlo simulation results for electric scooter GHG emissions per passenger-mile under different modeling assumptions (Hollingsworth *et al.*, 2019).

general flavors. The first is comprehensive models meant for in-depth analysis or research, where the user has control over each parameter and can build custom, complex models. The second is screening models meant for non-specialists that consist of pre-calculated impact results for common materials and processes. This is the approach taken by online calculators, product design and architectural software, and many sector-specific LCA tools.

Table 11.2 shows an excerpt of the Inventory of Carbon and Energy, with embodied energy and embodied carbon values per kg of selected building materials. All the user needs to do is figure out how much of each material is needed for the product in question, multiply by each respective factor, and sum to find the cradle-to-grave results. This approach is fast and easy if all one needs is results for reporting or the choice of materials is fixed, but understanding exactly where and why certain materials are causing impacts instead requires comprehensive LCA software.

LCA software packages are continuously being refined, as are the LCI data and LCIA methods that are loaded into them. When choosing an LCA software platform, be sure to check which versions of datasets and methods and being used, and that they are providing good coverage for the types of flow and environmental impacts that are part of your goal and scope.

Table 11.2.   Excerpt of pre-calculated inventory and impact factors of selected building materials from the free Inventory of Carbon and Energy (University of Bath, UK and Circular Ecology).

Material	Energy MJ per kg	Carbon kg CO_2 per kg
Aggregate	0.083	0.005
Aluminum (incl. 33% recycled)	155	8.24
Bitumen (general)	51	0.38–0.43
Bricks (common)	3	0.24
Concrete block (medium density)	0.67	0.07
Concrete (1:1.5:3)	1.11	0.16
Copper (incl. 37% recycled)	42	2.6
Glass	15	0.85
Gold	100,000	10,000
Iron (general)	25	1.91
Lead (incl. 61% recycled)	25.2	1.57
Plywood	15	1.07
Polyurethane insulation (rigid foam)	102	3.48
PVC (general)	77.2	2.41
Stainless steel	56.7	6.15
Steel (avg. recycled content)	20.1	1.37
Timber (excl. sequestration)	8.5	0.46
Vinyl flooring	65.6	2.92
Wool carpet	106	5.53

11.6   LCA Limitations and Purposes

LCA has enjoyed widespread popularity as an ideal way to quantify environmental impacts at a systems level, and it has been applied in thousands of studies. But despite its widespread use, it is important to keep in mind its limitations. Most generally, LCA is meant to provide information to support decisions. Its core competency is providing a broad life cycle perspective, but there are other perspectives and tools that can complement it, such as risk analysis, exposure assessment, or cost-benefit analysis.

As with any model, the quality of LCA results depend on the quality of the data inputs. LCI datasets continue to grow, but simply cannot keep pace with all of the millions of items that are produced worldwide, using a variety of processes and in different locations, all of which are continuously changing in technology and efficiency. In addition to which data sets to use, the LCA practitioner has to make various decisions about how to carry out the modeling, especially around the scope and setting of the study. As a result, LCAs by different assessment teams can produce different, though perhaps equally defendable, results. This is one reason why documenting assumptions in LCA is so important, as it allows readers to see what is driving differences among studies.

There are many efforts to harmonize LCA, including developing reference datasets of government data, more detailed technical guidance (such as the International Reference Life Cycle Data System), standards around how flows are named and classified, guidelines for specific types of products called Product Category Rules (PCRs), and others. These efforts are critically important for LCA to be used for absolute measurements (as in carbon accounting) and policy settings where consensus around a uniform approach is vital.

LCA methods also continue to evolve. One of the greatest opportunities is to improve the spatial and temporal resolution of both inventory data and impact assessment methods. Current datasets are often resolved to the national or sometimes regional scale, but in reality, the location and timing of emissions matters in terms of downwind or downstream exposures. Allowing LCIA methods to incorporate this type of information will bring LCA results more in line with the sophisticated environmental models that underpin policy, and will make results more useful to communities who have specific questions about their particular situation and for whom generic results are not that useful (see Hellweg and Milà i Canals in Further Reading). A second opportunity has to do with the mathematical structure. LCA is a linear model: double the flows and you get double the impacts. But for some types of environmental impacts, notably toxicity, responses can be nonlinear. Lower concentrations of a pollutant might have no effect (or even be essential to life). Capturing these threshold effects consistently across all types of flows has been difficult, but is an active area of research.

Despite these limitations, LCA remains an extremely useful and flexible tool, and a core one within industrial ecology. Its large community of practitioners and researchers continue to generate data and add new methods and capabilities, including the integration of important environmental and social impacts. This chapter has covered the basics but there are many resources available (some included in Further Reading) for those who want to learn more.

Further Reading

Bare, J.C. Tool for the Reduction and Assessment of Chemical and Other Environmental Impacts (TRACI), Version 2.1 — User's Manual; EPA/600/R-12/554, 2012.

Guinée, J. *et al.*, *Handbook on Life Cycle Assessment — Operational Guide to the ISO Standards*, Dordrecht, The Netherlands: Klewer Academic Publishers, 2002.

Heijungs, R., and S. Suh, *The Computational Structure of Life Cycle Assessment*, Springer Science & Business Media, 2002.

Hellweg, S., and L. Milà i Canals, Emerging approaches, challenges and opportunities in life cycle assessment, *Science*, *344*(6188), 1109–1113, 2014.

Hollingsworth, J., B. Copeland, and J.X. Johnson, Are e-scooters polluters? The environmental impacts of shared dockless electric scooters. *Environmental Research Letters*, *14*(8), 084031, 2019.

Matthews, S., C. Hendrickson, and D. Matthews. *LCA Textbook*. Online at: https://www.lcatextbook.com/

UNEP-SETAC Life Cycle Initiative. Online at: https://www.lifecycleinitiative.org/

Williams, E. D., R.U. Ayres, and M. Heller, The 1.7 kilogram microchip: Energy and material use in the production of semiconductor devices, *Environmental Science & Technology*, *36*(24), 5504–5510, 2002.

Chapter 12

Economics Tools: Techno-Economic Analysis, Life Cycle Costing, and Valuation

Chapter Concepts

- Industrial ecology often employs economic analyses to complement environmental and social sustainability tools.
- Techno-economic assessment (TEA) and life cycle costing (LCC) provide a life cycle approach to financial analyses of products, processes, and projects.
- Valuation can be used to convert environmental and social benefits impacts into economic terms in order to facilitate communication and comparison.

12.1 A Life Cycle Perspective on Costs

One of the main goals of industrial ecology is to apply a life cycle perspective to understanding physical products and systems, so that we can make design and consumer decisions that are holistically better for the environment. Tools like LCA can aid us in quantifying emissions and resources concerns and identify hotspots in a product life cycle that are targets for improvement. But of course, environmental performance is just one of many considerations that can inform decisions. Another vital consideration is economic costs and benefits.

In this chapter, we will briefly introduce several economics tools that are widely applied in industrial ecology, either in parallel to environmental analyses, or as integrated approaches. As with LCA, these tools apply a life cycle perspective to ensure that we are seeing the whole picture, and not making economic decisions now that shift costs to future generations. Techno-economic analysis (TEA) is a process-based modeling tool for quantifying the total costs of production, typically carried out from a manufacturing perspective when considering investment in a new product or process. The closely related tool of life cycle costing (LCC) has a wider scope, covering the entire life cycle costs of a physical asset like a bridge. Both TEA and LCC are akin to the inventory phase of LCA in that they map inputs and outputs for each step of a process, but account for flows of money rather than flows of mass or energy. Economic valuation, on the other hand, is akin to the impact assessment phase of LCA, but in units of economic damages or benefits rather than environmental and health consequences.

Economics tools are an important component of an industrial ecologist's toolbox, especially in interfacing with industry and policymakers. Modeling both environmental and economic impacts ensures that recommendations are more likely to produce win-win-win scenarios that are good for people, prosperity, and the planet.

12.2 Techno-Economic Analysis

As long as humans have been making products for barter or sale, there has been a calculus of whether it is worthwhile. The modern manifestation of that calculus is called techno-economic analysis (TEA), widely used by process and manufacturing engineers to quantify costs associated with production.

Costs fall into two major categories. Capital costs are those associated with the construction of the production facility, including construction materials, equipment, land, labor, as well as services such as financing, design, legal services, permitting, and insurance. Once the plant has been constructed and commissioned, operating costs are those incurred while the plant is running, including for material inputs, energy and water utilities, and labor. Adding capital and operating costs together gives an estimate of the cradle-to-gate costs of production, and therefore whether the process will be economically competitive.

But how to add capital and operating costs? A central tenet of economic analysis is the "time value of money", which holds that cash flows that occur at different times have different values. This is because money today can be placed into a risk-free investment that produces interest or has a rate of return. Thus, the future value FV of money spent today (present value PV) depends on the prevailing rate of return i and the number of compounding periods n into the future:

$$FV = PV \cdot (1+i)^n \qquad (12.1)$$

Conversely, the present value of money spent in the future can be found by rearranging the terms as

$$PV = \frac{FV}{(1+i)^n} \qquad (12.2)$$

Initial capital costs are assumed to be spent today (year 0) while operating costs (and later capital costs for multi-year construction projects) are spent in the future. So, in order to add them together, we discount costs in each future year over the plant lifetime N to their present values and add to capital costs in year 0 to obtain the net present value NPV:

$$NPV = \sum_{n=0}^{N} \frac{FV_n}{(1+i)^n} \qquad (12.3)$$

This is called a discounted cash flow analysis and is the basis of all manner of investment decisions. The net present value divided by the total production volume gives the unit cost of production, which in turn sets a minimum price for the product in order for the plant to be profitable. For fuels, this is called the minimum fuel selling price (MFSP, per liter or gallon); for electricity it is called the levelized cost of electricity (LCOE, per MWh or kWh).

The starting point of a TEA, just as in a process-based LCA described in the previous chapters, is a detailed process flow diagram (PFD). Figure 12.1 shows a PFD for a plant that produces styrene from

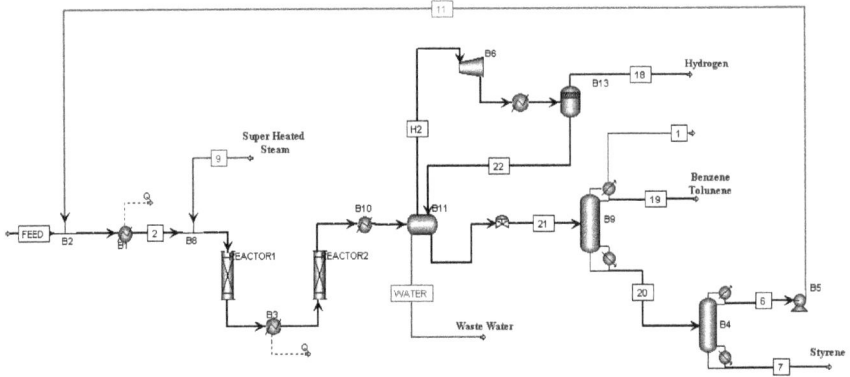

Figure 12.1. Process flow diagram (PFD) of styrene production from ethylbenzene modeled in AspenPlus (Parvatker and Eckelman, 2019).

dehydration of ethylbenzene. Specialized process simulation software has been developed that can then calculate the required size of each piece of equipment (for capital costs) and the material, energy, and water requirements (for operating costs). Labor and other operating costs are determined based on similar plants.

Because its process models are highly detailed, TEA can identify exactly which aspects of production most contribute to costs, and test how costs can be reduced through process modifications. TEA can also model how production costs will be influenced by external factors, for example if the price of a certain input falls or if a policy incentive like a production tax credit is introduced. Manufacturing operations can be designed to be more or less efficient; TEA is used as a process design tool to optimize designs based on profitability. For example, many chemical plants operate with a conversion efficiency for the feedstock of <100%. Higher efficiencies are physically possible (with additional separation and recycling stages), but may not be economically worthwhile.

12.3 Life Cycle Costing

While TEA focuses on cost estimation for process manufacturing, the related tool of LCC, life cycle cost analysis (LCCA), or total cost of ownership (TCO) considers ownership costs over the entire lifetime of a generic physical asset. TEA and LCC both quantify capital and operating costs and apply cash discounting methods in order to compare options and inform investment decisions; the main differences are the ownership perspective and the cradle-to-grave scope.

As an example of how LCC is used, consider the choice between two technology options for a home water heating system: a high-efficiency heat-pump system with substantial capital costs of $2,000 but low operating costs of $200 per year versus a less efficient natural gas fired system with lower capital costs of $800 but higher operating fuel costs of $600 per year. A typical water heater has a lifetime of 10 years. Table 12.1 shows a LCC discounted cash flow model for the comparison using a discount rate of 5% (and ignoring any salvage value or energy price inflation):

Even though the up-front (or "first") cost of the heat-pump system is 2.5 times that of the gas one, over the total lifetime of the water heater, the heat-pump system is 35% cheaper! Psychologically, humans tend to focus on first costs when making investment choices, so LCC is essential for prudent decision-making.

The example above is but one case of one of the most common questions in life cycle studies: is it worth it to invest more in the materials and

Table 12.1. Example LCC comparison table for home water heating systems.

Year	Heat-Pump nominal	Heat-Pump discounted	Gas nominal	Gas discounted
0	$2,000	$2,000	$800	$800
1	$200	$190	$600	$571
2	$200	$181	$600	$544
3	$200	$173	$600	$518
4	$200	$165	$600	$494
5	$200	$157	$600	$470
6	$200	$149	$600	$448
7	$200	$142	$600	$426
8	$200	$135	$600	$406
9	$200	$129	$600	$387
10	$200	$123	$600	$368
LCC		**$3,544**		**$5,433**

manufacturing stage in order to achieve benefits in the use stage? Similar LCC studies are routinely conducted to evaluate investments in green buildings or renovations, design and maintenance of infrastructure, or switching to a new technology such as electric vehicles. The very same core question motivates many LCA studies as well; literally thousands of LCA articles have been published on embodied versus operational carbon in buildings, for example in evaluating insulation products. By conducting LCA and LCC on the same system in parallel, industrial ecologists can provide additional value to consumers and policy-makers than by just considering environmental impacts.

12.4 Economic Valuation

So far we have learned about cost analysis in terms of actual costs that are borne by a manufacturing company or an asset owner. But there are others in the larger system who bear costs as well. If a manufacturing facility gives off air pollution that impacts the health of downwind residents, their hospital bills are certainly real, but they are not paid by the facility owner. Or if that pollution causes acid rain that kills a protected forest, no cash costs are incurred necessarily, but there is certainly a loss to society and to the natural world. These types of costs are called *externalities* — they are the costs that occur external to a process or asset that are not reflected in the financial decisions of firms or in the prices of goods and services. For example, a recent meta-analysis found that the mean externalities associated with coal-fired electricity generation is 14.5 ¢/kWh (Sovacool *et al.*, 2021), which is higher than the market price of electricity in many countries. Considering energy and transport, the authors estimate global externalities of nearly $25 trillion, or 29% of global GDP! Clearly these external costs are large and we should make every attempt to quantify and include them in decision-making. This process is called "economic valuation" and is one of the main objectives of the field of environmental economics.

Air pollution is an important source of externalities, but there are many others. If a road construction project causes traffic delays, externalities will include lost time, additional fuel costs, as well as air pollution impacts from idling vehicles. Residents who live nearby might also experience noise, light, and vibration disturbances during construction.

New construction in a coastal wetland area might negatively impact recreational activities like fishing. It might also degrade "ecosystem services" like erosion prevention that natural systems provide to us for free, so that we have to spend cash to build engineered systems like coastal barriers to do the same job.

Quantifying external costs is complicated and depends very much on which externalities are included. Some external costs like crop losses are straightforward to quantify because there are established market prices. Others such as health damages can be inferred from large data sets on hospital admissions. But what is the cost of losing something that has no monetized value, like a beautiful vista or a species that goes extinct?

Over the years, environmental economists have developed a wide range of tools to aid in economic valuation for different types of externalities. One important technique is hedonic price analysis, which uses the price of a proxy to infer intrinsic value, such as how home prices change based on proximity to a park or a metro station can be used to value green space or public transit. Another technique is willingness-to-pay (WTP), which asks people how much they would be willing to pay to avoid a loss. For example, economists have found that people are willing to pay between $5 and $100 to protect an individual species from extinction. Taken across millions of households, this means that the value of that species to humanity could be in the billions of dollars, such that it is worth funding conservation programs accordingly. These non-market valuation techniques don't capture the intrinsic value of species or natural systems independent of their perceived value to humans, but they are still important tools in capturing external costs.

Public agencies and large institutions like development banks now oftentimes require an analysis of environmental and health externalities for any new project and have published standard tables of external costs for economic valuation. Table 12.2 shows published values for transportation projects from recent guidelines in the US. We can see, for example, that the cost of particulate matter pollution per metric tonne is quite high, as this is the leading cause of human health damages from air pollution, causing millions of premature deaths annually worldwide. Especially with the publication of standard, consensus factors by public agencies and international institutions, social costs are increasingly being included in LCC studies.

Table 12.2.   External costs for US transportation projects (US DOT, 2021).

Externality	Published value (in 2019$)
Vehicle travel (average) per hr	$17.90
NO_x emissions per mg	$15,700
SO_2 emissions per mg	$40,400
$PM_{2.5}$ emissions per mg	$729,300
CO_2 emissions per mg	$50

The last value in Table 12.2 is of particular note. This is the external cost associated with carbon greenhouse gas emissions, what is called the "social cost of carbon" (SCC). Climate change will affect human health in various ways, from increased risk of natural disasters and heat waves to higher prevalence of vector-borne diseases and crop failures leading to malnutrition. Economists use climate models to see how these various effects may manifest and then assign social costs based on the number of people affected. As with the transportation example, SCC is now being included in investment decisions or is being levied as an actual cost, either internally or through carbon markets and government policies. By internalizing the cost externalities of greenhouse gas emissions, the world will have a stronger financial incentive to decarbonize and meet our collective climate goals.

The place where industrial ecologists most use economic valuation is in life cycle impact assessment, specifically in the calculation of endpoint characterization factors. Recall that these factors quantify environmental impacts in terms of damages, in contrast to midpoint characterization factors that quantify physical changes in the environment. The most common endpoint categories are health damages and species loss. By assigning monetary values from economic valuation studies to each of these categories, LCA results can be expressed in financial terms, which may add to their policy relevance and influence.

Further Reading

National Renewable Energy Laboratory (NREL), U.S. Department of Energy (DOE), *Techno-Economic Analysis*, https://www.nrel.gov/analysis/techno-economic.html, accessed January 4, 2022.

Parvatker, A.G., and M.J. Eckelman, Comparative evaluation of chemical life cycle inventory generation methods and implications for life cycle assessment results. *ACS Sustainable Chemistry & Engineering, 7*(1), 350–367, 2018.

Sovacool, B.K., J. Kim, and M. Yang, The hidden costs of energy and mobility: A global meta-analysis and research synthesis of electricity and transport externalities. *Energy Research & Social Science, 72*, 101885, 2021.

U.S. Department of Transportation (DOT), Benefit-Cost Analysis Guidance for Discretionary Grant Programs, (2021), https://www.transportation.gov/sites/dot.gov/files/2021-02/Benefit%20Cost%20Analysis%20Guidance%202021.pdf, accessed January 4, 2022.

U.S. Environmental Protection Agency (EPA), *Life Cycle Assessment and Cost Analysis of Distributed Mixed Wastewater & Graywater Treatment for Water Recycling in the Context of an Urban Case Study*, EPA/600/R-18/280, 2019.

Industrial Ecology in Application

Chapter 13

Sustainable Design Frameworks: Design for Environment, Cradle-to-Cradle, Principles of Green Chemistry and Green Engineering

---

Chapter Concepts

- Sustainable design frameworks provide general guidance for creating sustainable products and processes.
- Design software used by designers, engineers, and architects is increasingly incorporating environmental considerations to allow for rapid assessment and iteration to meet environmental targets.
- Heuristics and "rules-of-thumb" allow for widespread implementation of IE-informed recommendations, but should be periodically updated based on changing circumstances.

13.1 Sustainable Design Frameworks

Industrial ecology tools allow us to quantify industry's use of resources, identify environmental hotspots in the supply chain of a product, or compare product options over their life cycles. This is critical information in understanding *what* and *where* improvements can be made in a manufacturing process or product, but not *how* to make them. Designers and engineers need guidelines in order to innovate on the next generation of products and materials. IE has long been allied with various sustainable design frameworks that provide general heuristics or principles to guide practitioners. In this chapter we will review three important frameworks that have broad applicability: Design for Environment (DfE), Cradle-to-Cradle (C2C), and the Principles of Green Chemistry and Green Engineering.

13.2 Design for Environment

With growing environmental consciousness around the world in the 1970s, manufacturers began to incorporate environmental considerations into their designs. Energy efficiency became a desirable trait and designers began to screen out chemicals with noted toxic or ecotoxic effects. Eventually these design criteria became formalized under a generic framework called "DfE". Each product and sector has different requirements and environmental considerations, many of which have been further standardized into eco-labels and certifications, but some overall DfE guidelines that are widely applicable are presented below.

Product designers, engineers, and architects must balance multiple attributes of their designs. In addition to potential environmental considerations, a product must, of course, technically perform well to fulfil its desired function, and the designer must also think about aesthetics, ergonomics, durability, repairability, manufacturability/constructability, cost, regulatory compliance, supply chain logistics, compatibility with existing systems, and many other constraints. This is clearly a lot to keep track of and so designers have developed a large suite of tools and methods to aid in the design process, such as computer-aided design (CAD). The past decade has seen tremendous progress in integrating environmental considerations into such design tools so that DfE has become a core goal.

One of the most common manifestations of DfE in manufacturing is material selection guides. Often, multiple materials might be satisfactory for a given performance specification, and the designer can then choose which one has the lowest cost, most attractive look, or in our case, the best environmental profile. Another important DfE application is in guiding choices among chemical products and evaluating alternatives, with a focus on safety. For example, the US Environmental Protection Agency

DfE program began in the early 1990s and has evolved into the Safer Choice program. Products that carry the Safer Choice eco-label must have constituent chemicals reviewed for carcinogenicity, reproductive/developmental toxicity, ecotoxicity, and persistence in the environment. In addition, the product must meet performance, pH, low-VOC, and sustainable packaging standards. Many similar eco-labels and safer chemicals screening tools exist around the world in order to guide chemical formulators.

The most widespread and important legislative approach to track hazardous chemicals in commerce is the European Union's REACH (Registration, Evaluation, and Authorisation of Chemicals) program. This action, which came into force on June 1, 2007, requires that chemicals manufactured in quantities greater than one metric ton must be registered, those manufactured in quantities greater than one hundred metric tons must be evaluated for toxicity, and those of high concern (such as carcinogens or mutagens) must be specially authorized for use. Downstream users of the chemicals are involved as well as those who make the chemicals. REACH has made it much more difficult and expensive to utilize unauthorized materials and provides a substantial incentive for attempts to meet product technical specifications without utilizing problematic new materials or new material combinations.

Various authors have devised DfE principles for specific sectors; below, we provide a general list that has wide applicability; several principles have been discussed in detail in previous chapters. As in industrial ecology, DfE recognizes the importance of a life cycle perspective and the desire to avoid unintended consequences. Accordingly, general DfE principles can be organized according to life cycle stage.

Materials and Manufacturing
- Choose readily available materials, e.g., those that avoid issues of supply risk.
- Minimize or avoid hazardous substances both in the product itself and in the manufacturing process, and utilize closed loops for necessary but hazardous constituents.
- Minimize materials diversity.
- Minimize energy and resource consumption in the production phase.
- Design the manufacturing process to minimize material waste.

- Use structural features and high-quality materials to minimize product weight and improve material efficiency.
- Employ materials that are renewable, reusable, and recyclable and/or recycled, whenever possible.

Packaging is another important area of environmental design. IE research has shown that the environmental impacts of some products are dominated by packaging choice, and that single-use packaging makes up about 40% of the entire market for plastics.

- Minimize packaging materials by reducing redundant layers, size, and thickness.
- Use recycled content packaging materials.
- Label packaging materials so that they can be identified for recycling or composting.
- Design packaging in standard sizes or geometries that can be efficiency packed for transport.
- Avoid the use of packaging materials such as PVC that may have harmful emissions during incineration.

Use
- Design for energy and water use efficiency.
- Minimize the use of consumables, emissions, and material dissipation during use.
- Choose materials, surface treatments, and structural arrangements to inhibit dirt, wear, and corrosion, thereby enhancing product life.
- Design products for durability that matches its technological lifetime.
- Design products that can be maintained, repaired, upgraded, reconfigured, or refilled through accessibility, labeling, modules, breakage points, and manuals.

End-of-Life
- Design for disassembly by using as few joining elements as possible and emphasizing screws, snap fits, and geometric locking that can be readily undone.

- Avoid material combinations that force time-consuming and otherwise unnecessary separation at the recycling stage.
- Avoid designs that make it impossible to separate materials or lead to cross-contamination, thus inhibiting recyclability.
- Minimize painting and fillers.

In most situations, designers following these DfE principles will produce much "greener" products than would otherwise be the case and will contribute to the sustainability of their firm and their planet.

As mentioned above, numerous product-specific DfE lists have been developed across nearly every sector. Many of these DfE lists have been formalized into certifications, standards, or eco-labels. DfE lists of generally agreed-upon environmental criteria provide manufacturers with an easy-to-follow map to designing greener products. Products that become DfE certified have a market advantage, as industries and consumers use DfE eco-labels to make environmentally-conscious choices. DfE programs also bolster the reputation of companies and their environmental, social, and corporate governance (ESG), making them more attractive to investors and business partners. As such, DfE standards and labels are extremely popular, with hundreds of programs currently running around the world.

13.3 Cradle-to-Cradle

In the LCA section (Chapters 10 and 11), we learned about different scopes and life cycle phases, including end-of-life. While reuse and recycling are certainly included in a cradle-to-grave LCA, the terminology implies that, eventually, materials and products come to the end of their life cycle. In the early 2000s, German chemist Michael Braungart and American architect William McDonough came together to espouse an approach they called "C2C", which aims to mimic nature by eliminating the very notion of waste and end-of-life. The goal of C2C is to design products and systems that can be infinitely cycled, either as biological nutrients that can be returned to nature, or as technical nutrients that be recycled repeatedly without being "downcycled" to products of lesser quality. By closing nutrient loops, all materials eventually return to their cradle to be remade into new products.

The C2C design framework shares much of its vision with industrial ecology, especially the notions of "waste = food" and closing material loops as much as possible. But where IE largely studies the interplay of production, consumption, and technology as it exists today, C2C is about designing the products and systems of the future. Where LCA might be used to choose the least bad option (i.e., with lowest environmental impact) or what is sometimes called "eco-efficiency", C2C envisions products that are instead "eco-effective", with designs focused on creating environmental benefits. C2C designs are renewable and rejuvenating, rather than depleting; for example, green buildings treat water, purify air, and provide attractive spaces for people to come together to work, rather than simply being more energy efficient. The C2C framework reminds us that it is important to focus on the positive power of technology and design and not just on negative consequences. Accordingly, a new approach in LCA put forward by Greg Norris and colleagues is called product "handprinting" (as opposed to footprinting), which quantifies the net environmental *benefits* that a product has over its life cycle.

The C2C design framework is stricter than many of the DfE principles listed in the previous section, and there is some criticism that it cannot realistically be applied to many categories of products and is more inspirational than practical. Nevertheless, a certification scheme has been created based on the C2C framework and numerous products have gone through the certification process, which sets required performance tiers across the five areas of material health, product circularity, clean air & climate protection, water & soil stewardship, and social fairness (see Further Reading).

13.4 Principles of Green Chemistry and Green Engineering

A major theme of green design frameworks is chemical toxicity and product safety. As with the C2C framework, we must not only avoid substances of concern, but also innovate to create new substances that have the functional properties we want but without negative side effects. The most influential paradigm in this area is the 12 Principles of Green Chemistry (Table 13.1), created by chemists Paul Anastas and John Warner in the late 1990s. Their work launched the field of green chemistry, now a worldwide community of academic and industrial chemists with hundreds of academic programs.

Table 13.1. 12 Principles of Green Chemistry (Anastas and Warner, 1998).

1. It is better to prevent waste than to treat or clean up waste after it has been created.

2. Synthetic methods should be designed to maximize incorporation of all materials used in the process into the final product.

3. Wherever practicable, synthetic methods should be designed to use and generate substances that possess little or no toxicity to human health and the environment.

4. Chemical products should be designed to preserve efficacy of function while reducing toxicity.

5. The use of auxiliary substances (e.g., solvents, separation agents, etc.) should be made unnecessary wherever possible, and innocuous when used.

6. Energy requirements should be recognized for their environmental and economic impacts and should be minimized. Synthetic methods should be conducted at ambient temperature and pressure.

7. A raw material or feedstock should be renewable rather than depleting whenever technically and economically practicable.

8. Unnecessary derivatization (use of blocking groups, protection/deprotection, temporary modification of physical/chemical processes) should be minimized or avoided if possible, because such steps require additional reagents and can generate waste.

9. Catalytic reagents (as selective as possible) are superior to stoichiometric reagents.

10. Chemical products should be designed so that at the end of their function they break down into innocuous degradation products and do not persist in the environment.

11. Analytical methodologies need to be further developed to allow for real-time, in-process monitoring and control prior to the formation of hazardous substances.

12. Substances and the form of a substance used in a chemical process should be chosen to minimize the potential for chemical accidents, including releases, explosions, and fires.

The Principles of Green Chemistry are just that: principles that can guide chemists in designing new molecules and syntheses. However, most by themselves do not convey information about how well a certain chemical or process performs. To operationalize the Principles, various metrics have been proposed that can track performance quantitatively. For example, one way to track Principle 2 is the metric of *Atom Economy*, defined as the percentage of reagents that make it into the final product. Increasing Atom Economy means that fewer reagents are required per unit of target chemical and that less chemical waste is produced in the process. However, chemical reactions involve more than just reagents; solvents, catalysts, buffers, and other auxiliary inputs are almost always involved. So, a broader metric was developed called the *Process Mass Intensity* (PMI), defined as the percentage of all chemicals involved in a process or step that make it into the final product. An ideal process that green chemists should strive for is when PMI is equal to 100% and there is perfect conversion.

From the perspective of Life Cycle Assessment, a mass-based metric like PMI is unsatisfying because it does not differentiate process chemicals in terms of their environmental impacts; essentially, all process chemicals are equally important in the metric. But it is important to remember that green design frameworks strive to provide simple heuristics to guide practice. Other tools such as green solvent selection guides can be used in concert to incorporate other life cycle considerations such as embodied emissions or toxicity.

Following the success of the Green Chemistry Principles, a parallel framework was developed for engineering called the 12 Principles of Green Engineering by Paul Anastas and Julie Zimmerman (Table 13.2).

The Principles of Green Engineering are broadly applicable to product engineering and design, and some are closely related to the DfE criteria described above. Together, the Principles of Green Chemistry and Green Engineering have been taken up and promulgated by various groups, including the American Chemical Society. Unlike other design frameworks that have been developed and commercialized into certifications or eco-labels, the Principles maintain their status as a set of guidelines, where chemists and engineers at companies strive to follow those Principles and associated metrics that best map onto their situations.

Table 13.2. 12 Principles of Green Engineering (Anastas and Zimmerman, 2003).

1. Designers need to strive to ensure that all materials and energy inputs and outputs are as inherently nonhazardous as possible.

2. It is better to prevent waste than to treat or clean up waste after it is formed.

3. Separation and purification operations should be designed to minimize energy consumption and materials use.

4. Products, processes, and systems should be designed to maximize mass, energy, space, and time efficiency.

5. Products, processes, and systems should be "output pulled" rather than "input pushed" through the use of energy and materials.

6. Embedded entropy and complexity must be viewed as an investment when making design choices on recycle, reuse, or beneficial disposition.

7. Targeted durability, not immortality, should be a design goal.

8. Design for unnecessary capacity or capability (e.g., "one size fits all") solutions should be considered a design flaw.

9. Material diversity in multicomponent products should be minimized to promote disassembly and value retention.

10. Design of products, processes, and systems must include integration and interconnectivity with available energy and materials flows.

11. Products, processes, and systems should be designed for performance in a commercial "afterlife".

12. Material and energy inputs should be renewable rather than depleting.

13.5 A Cautionary Note on Design Guidelines, Heuristics, and Principles

Design frameworks and the heuristics and principles that comprise them are so useful because they provide general guidance, saving time and effort. But it is important not to be complacent and over-generalize. There are so many different types and uses of products in the world that there are bound to be exceptions to each framework. For example, bio-based plastics may not be environmentally preferable to petrochemical plastics, depending on the particular feedstock, where it was cultivated, and the processing route that is used. IE tools, especially LCA, should still be used in concert with sustainable design frameworks to ensure that new products and materials do in fact have more sustainable outcomes for their specific use case.

We live in a rapidly changing world. What was environmentally preferable five years ago may not be so five years from now, and vice-versa. Sustainable design frameworks, eco-labels, and sustainability certifications must be updated regularly, informed by a life cycle and systems perspective, so that they continue to point in the correct direction.

Further Reading

Anastas, P.T., and Warner, J.C. *Green Chemistry: Theory and Practice*; Oxford University Press; New York, 1998.

Anastas, Paul T., and Julie B. Zimmerman, Design through the 12 principles of green engineering, *Environmental Science & Technology, 37*(5), 94A–101A, 2003.

Ashby, M.F., *Materials and the Environment: Eco-informed Material Choice*. Elsevier, 2012.

Cradle to Cradle Products Innovation Institute, 'Introducing the Cradle to Cradle Certified Product Standard Version 4.0', https://www.c2ccertified.org/get-certified/cradle-to-cradle-certified-version-4, accessed December 28, 2021.

McDonough, W., M. Braungart, P.T. Anastas, and Zimmerman, J.B., Applying the principles of green engineering to cradle-to-cradle design, *Environmental Science & Technology, 37*(23), 434A–441A, 2003.

Norris, G.A., J. Burek, E.A. Moore, R.E. Kirchain, and Gregory, J., Sustainability Health Initiative for NetPositive Enterprise handprint methodological framework, *The International Journal of Life Cycle Assessment, 26*(3), 528–542, 2021.

US Environmental Protection Agency, 'Safer Choice', https://www.epa.gov/saferchoice, accessed December 28, 2021.

Zimmerman, J.B., P.T. Anastas, H.C. Erythropel, and Leitner, W., Designing for a green chemistry future, *Science, 367*(6476), 397–400, 2020.

Chapter 14

Remanufacturing

Chapter Concepts

- The remanufacturing of products preserves the productive life of materials and products, saves energy, and minimizes waste.
- Product remanufacturing is aided by close links between the original manufacturer and the remanufacturer, links that have often not been well established.
- The business of remanufacturing, despite its materials and environmental benefits, is quite challenging and is, in general, poorly supported by government policy.

14.1 The Remanufacturing Concept

The typical sequence for the manufacture and use of a technological product is pictured in Figure 14.1. The process begins with extraction of raw materials, transformation into usable forms of the materials, fabrication of components, and product assembly. After distribution and use, the owner discards the product, thus meeting the traditional sequence of "make it, use it, and discard it when it is no longer useful". Some of the discarded materials might be recovered and reused. A more extensive approach, termed *remanufacture*, is to disassemble the product, clean it, replace any components as appropriate, upgrade the product to at least the original level of performance, refinish or repaint as appropriate, and return the product to service. This process is the subject of the present chapter.

Why might one wish to go through the remanufacturing process as opposed to acquiring a replacement for a product that is no longer satisfactory for one reason or another? One reason that has resonance for industrial ecologists is that remanufacturing retains 80–90% of the materials in the product for another use, thus avoiding the not uncommon practice of extracting materials, processing them, using them once, and then disposing of them. A second reason is that the remanufacturing process typically retains 60%–70% of the embedded energy in the original product. While these reasons seem enough to justify routine remanufacture, it turns out nonetheless that technical, operational, and economic factors can render remanufacture problematic.

14.2 The Remanufacturing Process

Remanufacturing begins with a recovered product, termed a *core*, which is "a previously sold, worn, or non-functional product or part, intended for the remanufacturing process". During reverse logistics, a core is protected, handled, and identified for remanufacturing to avoid damage and to preserve its value. A core is not waste or scrap, but it is not intended to be reused until it has been remanufactured (European Remanufacturing Network, 2020). There are two unstated implications in this definition. The first is that the core must be perceived by its owner to have value even when obsolescent. The second is that the owner must know how to access a reverse logistics network and transfer the core to that network. In practice, these two aspects require almost all remanufacturing to involve the large-scale industrial organizations who own the core or know who does and thus have substantial knowledge of the core design, manufacture, and current locale. This is the case, in fact, with the most commonly remanufactured products: automobile components, heavy duty equipment, aerospace products, medical devices, machinery, photocopiers, and the like.

Despite the perceived virtues of remanufacture, some product types are not suitable targets for this process for one or more of the following reasons:

- The product technology is not stable from one product cycle to the next (e.g., new medical devices are so different from previous devices as to make remanufacturing impractical).
- The process technology is not stable from one product cycle to the next.
- The product typically fails by dissolution or dissipation rather than functionally (e.g., brake linings).
- Critical components typically fail before the product itself fails.
- The cost to remanufacture is too high to justify the process.

Some of these challenges can be surmounted by more thoughtful design, as described below; others cannot be readily overcome.

Remanufacturing, as with initial manufacturing, is a multi-step process, as shown in Figure 14.1. The first stage, disassembly, immediately

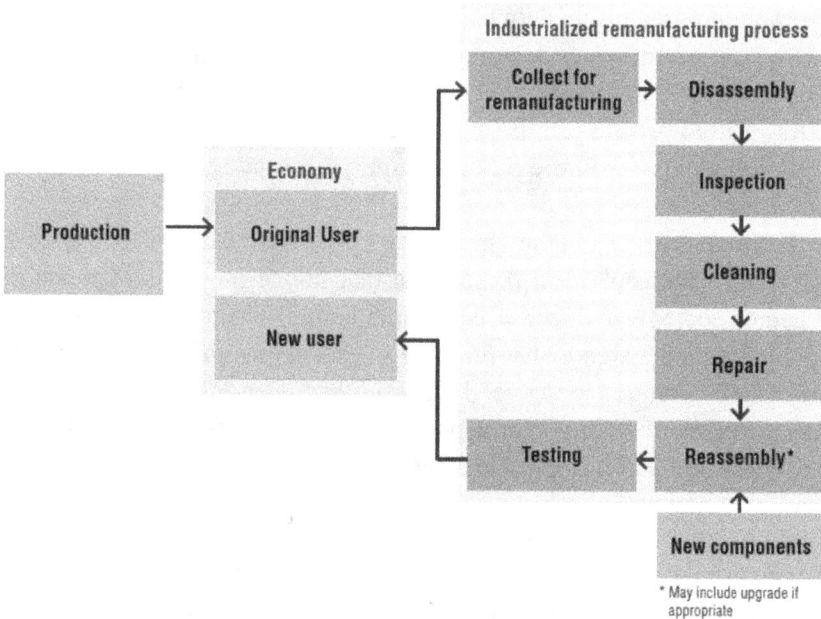

Figure 14.1.   The detailed remanufacturing flow chart (Nasr *et al.*, 2018).

presents the case for "design for remanufacturing", because remanufacturing is unlikely to occur unless it is relatively straightforward to disassemble the core with tools routinely available, and to do so while doing no damage. The core, and perhaps its subparts, are then inspected and cleaned to the level typical of new components. If repair is needed the core is taken to the new component level. It is then reassembled and tested to make sure that the remanufactured product is the equivalent of the original in appearance and performance.

An alternative that can be useful for some cores, such as office furniture suitable for reuse, may be to remanufacture the product into a form useful to consumers seeking a lower but still useful level of performance. This creates the "cascade loop", which retains the product and its materials for at least one additional product cycle before remanufacturing becomes impractical or too expensive to justify.

14.3 Product Evolution During Remanufacturing

The rapid progress of modern technology enables manufacturers to incorporate new performance features into their products at an increasingly rapid rate. As a consequence, many products become obsolete quickly. Unless a used product can be upgraded during remanufacturing, there is little opportunity for future product employment rather than scrappage at end of life.

It is interesting from the "suitability for remanufacturing" perspective to return to the ecological perspective that lies at the basis of much of industrial ecology, and to ask why some natural organisms evolve more rapidly than others, and what properties (phenotypes) of those organisms make such evolution possible. This issue has been termed *evolvability*, a characteristic that favors three phenotypic characteristics that influence variation (see Calcott in Further Reading):

- Minimize disadvantages resulting from changes.
- Increase the variety of results enabled by the changes.
- Reduce the number of changes needed to evolve into a more useful form.

Calcott suggests that the application of these concepts in engineering product design and product manufacture, as with natural selection, has the potential to generate more highly evolvable and useful technological artifacts. What might be some of the characteristics of the products that could be produced by such a procedure? The answer to this question can perhaps be best illustrated by the example of the Xerox photocopy machine, probably the first product for which the design team seriously considered remanufacture. The team first determined the aspects of the design that were likely to remain relatively stable over future generations of the product, determining that the size of the frame (i.e., the overall product dimensions) and the electrical interface, among other attributes, fell into that category. They next asked what aspects of the product might well evolve, that is, what might be the transformative features of the photocopier of the future. The list included:

- Improved optics and electronics for better image scanning.
- Transformed image generation (better inks? electronic paper?).
- Improved paper handling (higher speed, fewer paper jams, etc.).

Given that foundational thinking, the copiers were then designed to be capable of evolving during remanufacture, especially as regards those features expected to significantly change over time and to do so as new features were introduced to subsequent designs. The anticipated transformable parts were made readily accessible, readily removable, and readily replaceable. Following this design philosophy over decades has enabled Xerox to maintain photocopy systems for as many as seven remanufacturing cycles.

The message for remanufacturing is obvious: anticipate product evolution and then design today's systems to facilitate upgrading them during future remanufacturing.

14.4 The Business of Remanufacturing

Despite the obvious benefits of remanufacturing — preserving the productive life and embedded energy of materials and products, minimizing waste, capturing critical materials that would otherwise be lost, etc. — the remanufacturing business is quite challenging. A list of the characteristics of the business includes:

- Increasing the first lifetimes of many new products renders core designs and manufacturing approaches often a distant memory, the need for replacement parts is often smaller than it had been historically, and the time to initial remanufacture is longer.
- Parts needed for remanufacturing are often not available if design evolution is very rapid.
- Electrical components are ever more numerous and are increasingly difficult to locate, identify, evaluate, and replace.
- The increasing complexity of products makes them challenging to remanufacture.
- Cores that have not been handled carefully after they leave service may require major initial processing because of contamination, grease, rust, or other impediments.
- A remanufacturing business that is not highly specialized (e.g., one that only remanufactures aircraft engines) must deal with a very wide range of products, including appliances, electronic and network equipment, heavy duty components (earth movers, mining equipment), machine tools, medical equipment, military defense platforms (armored vehicles, etc.), and more. Accordingly, a large and diverse stock of parts must be kept on hand.
- The financial picture for a remanufacturer can be very challenging because of such issues as legal constraints, the modest size of the market for remanufactured goods, warranties on some products, and the supply chain for recovered products suitable for remanufacturing.

Notwithstanding these challenges, remanufacturers can provide real service not only to society but also to original product manufacturers. Remanufacturers often know more about how equipment performs over time – how equipment tends to be used, what parts tend to fail and when, how amenable particular equipment is to multiple cycle of reuse, and the like. As with recycling in general, increasingly close links between the original product designer and the remanufacturer are likely to benefit all parties and to better preserve products and their constituents for multiple uses over time.

Remanufacturing can be aided by a governmental approach termed extended producer responsibility (EPR), the requirement that manufacturers take responsibility for their products at the end of useful life. The intent is that product reuse and recycling (and ultimately product design) will be much more likely to occur in environmentally responsible ways if the original manufacturer bears the burden. This approach has been adopted largely in Europe and has often involved industry-collaborative programs for product collection and subsequent treatment.

14.5 Linking Remanufacturing and Recycling

It is worth noting that Figure 14.1 implies that recycling is an option at end of product life. It is significant that if recycling occurs it implies that the next step will occur at the material transformation stage rather than at the components and fabrication stage, that is, that it will be elements, not products or product modules, that will be recycled. Consequently, the embedded energy in the assembled components will be lost even if recycling is perfect. And, as noted in Chapter 7, end-of-life recycling rates for many materials are very small, so the recycling and reuse of end-of-life materials is problematic. Therefore, from both a materials and energy standpoint, remanufacturing to the degree possible is likely to be far preferable to recycling at the end of a product's first (or even second or third) life.

Nonetheless, for situations in which remanufacturing is not the ideal approach, such as major design upgrades that make it unlikely that a remanufactured product would be competitive in the marketplace, design choices focused on recycling may be desirable. As an example, Ford Motor Company chose for its F-150 trucks a particular set of aluminum alloys designed to optimize recycling and avoid problems with contaminating "tramp" elements that made recycling difficult or impossible (Figure 14.2).

Remanufacturing has, in general, not been well aided by government policies and regulations. An important future job for governments and

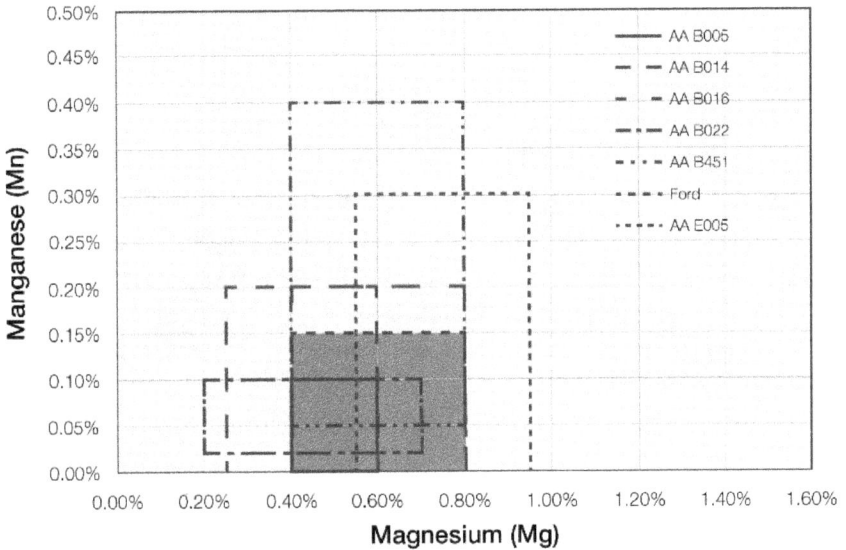

Figure 14.2.   Aluminum alloy selection for Ford F-150 trucks (Chappuis, 2015).

manufacturers is to support the remanufacturing industry so as to fully realize its benefits for society and the environment.

Further Reading

Allwood, J.M., Squaring the circular economy, in *Handbook of Recycling*, E. Worrell and M.A. Reuter, eds. Amsterdam: Elsevier, 2014, pp. 445–477.

Apple Corporation, Apple introduces Daisy, a new robot that disassembles iPhone to recover valuable materials, https://www.youtube.com/watch?v= 2Bu-gl7v-P8, accessed July 7, 2020.

Calcott, B., Engineering and evolvability, *Biology and Philosophy*, *29*, 293–313, 2014.

Chappuis, L.B., Material Specifications & Recycling for the 2015 Ford F-150, paper presented at *3rd Annual Global Automotive Lightweight Materials Conference*, Detroit, MI, 2015.

Cooper, D.R. and T.G. Gutowski, The environmental impact of reuse, *Journal of Industrial Ecology*, *21*(1), 38-56, 2015.

European Remanufacturing Network, Defining remanufacturing, https://www. remanufacturing.eu, accessed August 10, 2020.

Klausner, M., W.M. Grimm, and C. Hendrickson, Reuse of electric motors in consumer products, *Journal of Industrial Ecology*, *21*(1), 38–56, 2015.

Matsumoto, M., S. Yang, K. Martinson, and Y. Kainuma, Trends and research challenges in remanufacturing, *International Journal of Precision Engineering and Manufacturing-Green Technology*, *3*(1), 129–142, 2016.

Nasr, N. *et al.*, *Re-defining Value — The Manufacturing Revolution. Reuse, Refurbishment, Repair, and Direct Reuse in the Circular Economy*, Paris: International Resource Panel, ISBN 978-92-807-3720-2, 2018.

Rahito, D.A.W. and A.H. Azman, Additive manufacturing for repair and restoration in remanufacturing: An overview from object design and systems perspectives, *Processes*, *7*(11), 802, 2019.

Tolio, T. *et al.*, Design, management and control of demanufacturing and remanufacturing systems, *CIRP Annals — Manufacturing Technology, 66*, 585–609, 2017.

Chapter 15

Industrial Ecology and Energy

Chapter Concepts

- The availability and quality of energy available to humans has been a key factor in the development of our civilization.
- Energy use is a prerequisite for extracting, processing, transporting, and using materials, the sum of which is called embodied energy.
- The embodied energy of metals, chemicals, plastics, and fibers are currently all significantly linked to greenhouse gas emissions, and a rapid transition to renewable energy sources for material production is critical to mitigate climate change.
- Complex and globally-integrated supply chains for major technological products complicate attempts to minimize the embodied energy for many manufactured products.

15.1 Energy and Societal Development

Transformation of a material or a product from one state to another cannot happen without a source of energy. Indeed, it is not an exaggeration to say that a transition from one source of energy to an improved source has empowered humans to make great leaps in their own comfort, capabilities, and accomplishments, over and over again in history. The first energy source was fire, which along with the cooking of food also enabled the clearing of land for crop production and, later, the processing of small amounts of metal ores into tools and weapons. Waterpower, when harnessed from nearly rivers, made possible mills for the production of flour and also many early industrial processes such as the manufacture of fabrics. Crude iron was made by utilizing charcoal from wood to heat ores and generate impure iron capable of being worked into tools. The next big jump was the discovery of coal deposits and their use in enabling the growth of embryonic industries in the 17th century CE. With the advent of oil drilling two centuries later, the Industrial Revolution transformed largely agricultural societies into the industrialized world we know today.

Energy use falls into two categories: it can be used in a way that is productive over the long term in the creation of materials and products, such as the manufacture of tools or housing. This is called *embodied energy.* Or, it can be used directly in the short term to provide energy services and then dissipated, as in the heating of buildings for the comfort of occupants. As is well known, the widespread use of coal and oil for both productive and dissipative purposes resulted in environmental problems as it enabled technological and societal advances. Earth is now undergoing a massive shift in climate, with resulting changes in temperature, seawater level increases, ecosystem transformations, and numerous other transitions. The question now is whether transitioning away from fossil fuels and towards an energy system founded in renewable energy can minimize ecosystem impacts and enable a more stable and sustainable society can emerge in a time of rapid planetary evolution.

15.2 Global Energy Use for Materials

Many textbooks exist that discuss energy systems and technologies; here we focus on how energy flows interact with material flows to provide services to humanity. As with most issues in industrial ecology, quantification related to the use and loss of resources is essential for detailed understanding. We have already made the case that some of the resources that the planet provides are scarcer than others. This fact suggests that more energy is required to access scarcer materials than those materials present in higher concentrations or more convenient repositories.

The embodied energy of producing different metals is shown in Figure 15.1(a). The concept of embodied energy or cumulative energy demand should be familiar from the introduction to life cycle assessment in Chapters 10 and 11; LCI databases contain embodied energy values for a wide range of materials. Rhodium (Rh) has the largest embodied

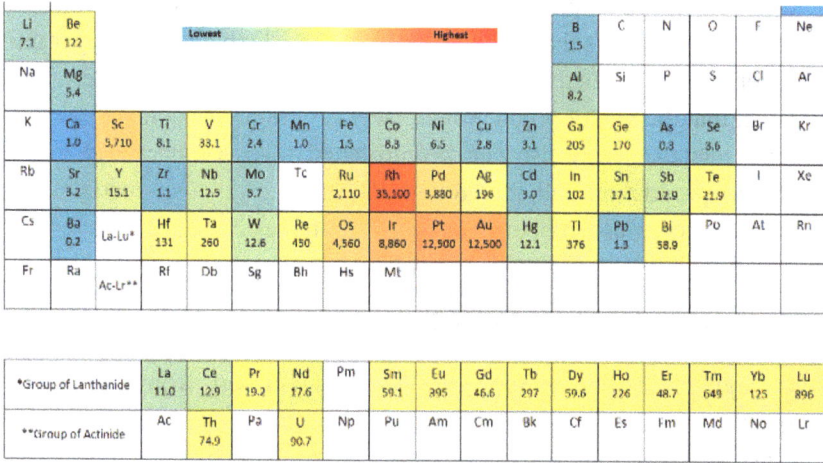

(a)

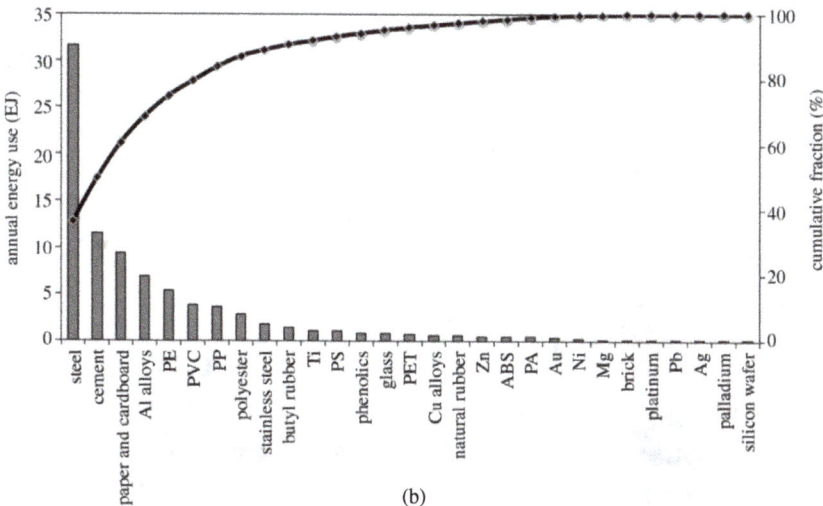

(b)

Figure 15.1. (a) Embodied energies in the elements of the periodic table (Nuss and Eckelman, 2014). (b) Annual energy use required by the demand of different widely-used materials (Gutowski *et al.*, 2012).

energy (more than 35,000 MJ/kg). The other "platinum group" metals (Pt, Ir, Os, Pd, and Ru) have very high embodied energies as well, as does gold (Au), largely due to their low concentrations in deposits. In contrast, many of the commonly-used metals (third row center) have quite low embodied energies.

An overview of the energy use attributable to the acquisition and processing of materials is that of Figure 15.1(b), in which energy use per unit of material is multiplied by the societal rates of use of those materials. It turns out that the rate of production of steel is so high that some 35% of all material-related energy use can be attributed to it. Cement and paper/cardboard production call for around 10% each. All other materials — metals, plastics, glass, rubber, *etc.* constitute the rest.

15.3 Energy Use by Major Industrial Sector

The major industrial sectors can be distinguished from an energy perspective by tracking their energy use over time, as shown for the iron and steel sector in Figure 15.2. When all materials sectors are considered, metal acquisition and processing consumed nearly a quarter of all

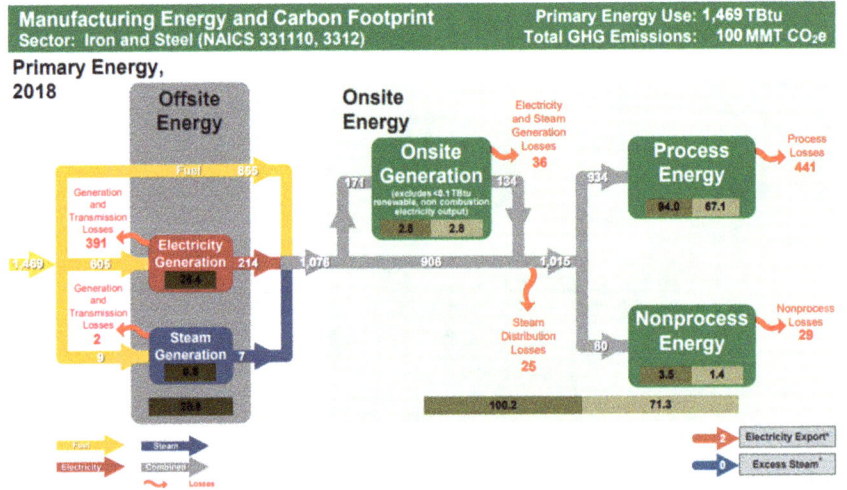

Figure 15.2.   Process energy flows in the iron and steel sector of the United States (U.S. Department of Energy, 2021).

industrial energy. Cement at about 13% was next, with chemicals (including plastics) at 8%. "Other industries" constituted more than a third of all industrial energy use in activities such as transforming materials, machining and processing them, and transporting the final products to customers. Wastewater treatment and landfills, industries that deal with discards and methane emissions from those discards, consume a surprising 15% of industrial energy use.

Once given a quantitative perspective on energy and emissions from a particular process or energy sector, further analysis can demonstrate opportunities for reducing energy demand. An example of such an exercise by the US Department of Energy (DOE) is shown in Figure 15.3, which addresses

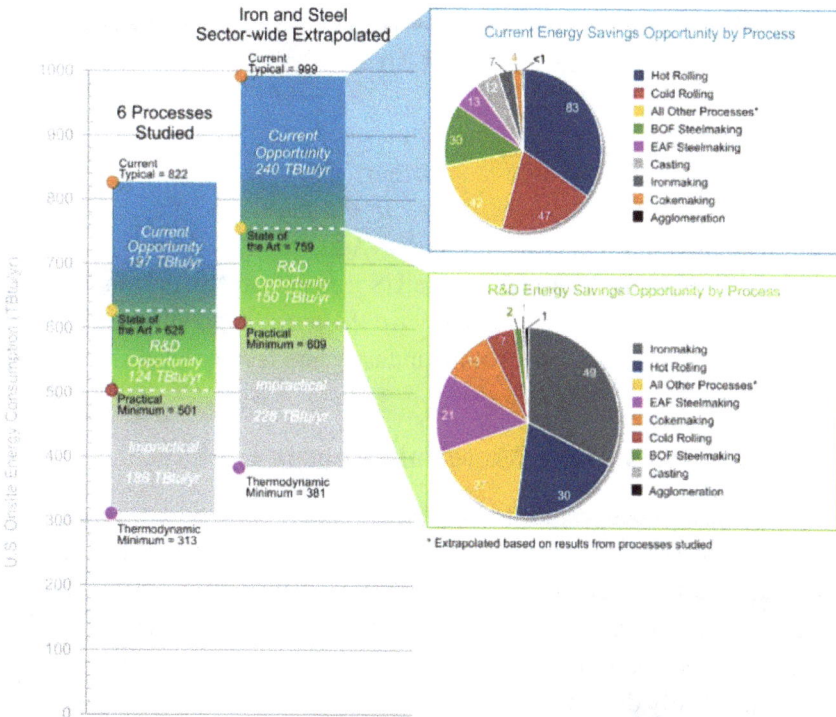

Figure 15.3. Distributions of energy in the iron and steel sector of the United States. Energy saving opportunities in the iron and steel sector of the United States. Savings that could be realized are shown in blue, potential energy saving opportunities that require research and development to realize appear in green, minimum energies needed for the processes are shown in gray, thermodynamic minima are noted in purple (U.S. Department of Energy, 2015).

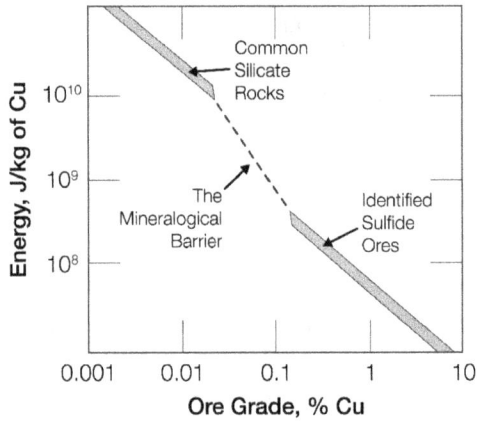

Figure 15.4. The energy necessary to recover copper from its ores and by atomic substitution from common silicate materials (Gordon *et al.*, 1987).

energy use in the US iron and steel industries. (DOE has also completed similar studies for other industrial sectors, including aluminum, plastics, and carbon-fiber reinforced polymers.) In the figure, the gray, green, and blue areas represent the ranges of energy use resulting from process and operational differences throughout the US iron and steel industrial sectors. Analyses done to this level of depth have the potential to stimulate investments in process improvement at the industry and factory levels.

15.4 Dynamics in the Materials-Energy Linkage

The relationships between materials and the energy needed for their acquisition and processing are not static, but evolve with the sources and types of materials and with technology development. A well-studied example is that of the energy required for the extraction and processing of metal ores. The ore grades (percent of the metal of interest in a typical ore deposit) for most metals are clearly decreasing over time as richer deposits are mined out and poorer deposits are then exploited. It appears that a logarithmic relationship exists between ore grade and required energy (Figure 15.4), and thus the amount of energy required to access the same quantity of extracted metal is very likely to increase dramatically. Over time, new approaches to material use, vastly improved technologies for materials processing, and/or new energy/materials relationships may be required to simultaneously provide the material resources for humanity as well as to maintain a sustainable global environment in which to enjoy the fruits of those resources.

15.5 Energy-Materials-Society Challenges

Although the relationship between the employment of materials in various ways and for various purposes, together with the energy thereby required, is quite likely to constitute a challenge in that the environmental implications are directly related to bettering the human condition in one way or another. An obvious approach to this challenge is to provide the energy needed to serve humanity from renewable resources, or perhaps with some use of carbon capture and storage. Energy is clearly needed if society is to function, and thoughtful provisioning of that energy will therefore be clearly needed as well.

Further Reading

Gordon, R.B., T. Koopmans, W.D. Nordhaus, and B.J. Skinner, *Toward a New Iron Age?* Harvard University Press, 1987.

Gutowski, T.G., S. Sahni, J.M. Allwood, M. F. Ashby, and E. Worrell, The energy required to produce materials: constraints on energy-intensity improvements, parameters of demand, *Philosophical Transactions of the Royal Society A, 371*, 20120003, 2012.

Nuss, P., and M.J. Eckelman, Life cycle assessment of metals: A scientific synthesis, *PLOS One, 9(7)*, e101298, 2014.

Oberle, B. *et al.*, Global Resources Outlook 2019: *Natural Resources for the Future We Want*, Paris: United Nations Environment, ISBN 978-92-807-3741-7, 2019.

Rötzer, N., and M. Schmidt, Historical, current, and future energy demand from global copper production and its impact on climate change, *Resources*, 9, 44, 2019.

U.S. Department of Energy, Bandwidth Study on Energy and Potential Energy Savings Opportunities in U.S. Iron and Steel Manufacturing, June, 2015.

U.S. Department of Energy, Manufacturing Energy and Carbon Footprints (2018 MECS), December, 2021. https://www.energy.gov/sites/default/files/2021-12/2018_mecs_iron_steel_energy_carbon_footprint.pdf.

Chapter 16

Material Criticality

Chapter Concepts

- Critical materials are those with limited or restricted supply that are nonetheless embodied in products deemed to be of vital importance to industries or governments.
- Depending on the corporate or government entity conducting the evaluation, as many as half of the elements employed by modern technology can be designated as critical.
- Given the broad palette of materials in many of today's commercial products, avoiding the use of critical materials is increasingly difficult.

16.1 The Criticality Concept

In 2008 a committee of the US National Academies published an analysis that identified which non-fuel minerals might be termed "critical" to the national economy. In that work, eleven minerals or mineral groups were singled out as having at least some degree of criticality. To carry out its analysis, the committee devised a two-axis rating system (shown in Figure 16.1) based on "supply risk" and "impact of supply restriction".

The idea was that if a country, a region, or the world was unable to procure a specific material that was needed for an important technology or application, either because the supply was shrinking for some reason or

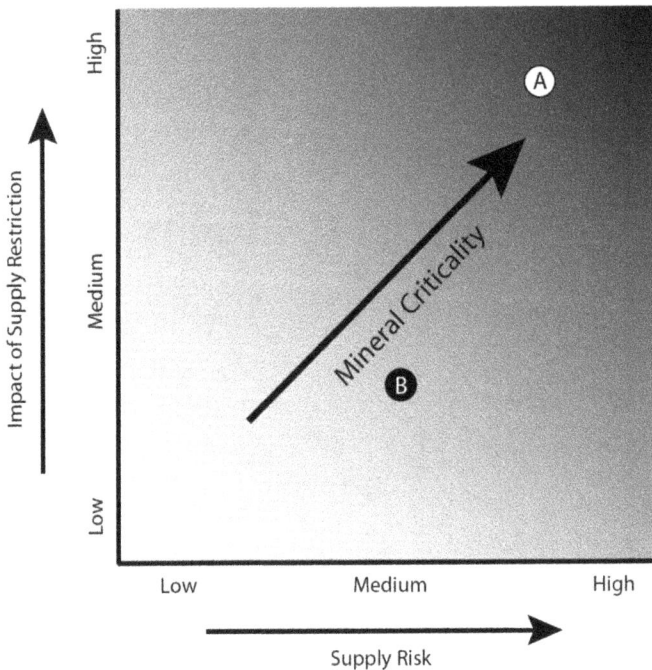

Figure 16.1. The criticality matrix originated by the US National Research Council. The degree of criticality increases as one moves from the lower left to the upper right corner of the figure. In this example, mineral A is more critical than mineral B (U.S. National Research Council, 2008).

that demand was rapidly growing as new applications emerged, the inadequacy could hamper industrial progress or endanger security. The second axis in Figure 16.1 was designed to evaluate the degree to which a supply-demand imbalance would cause disruption in industrial activities.

The National Research Council report generated short lists of "dimensions of primary availability" (for virgin materials) and "dimensions of secondary availability" (for previously used materials), as follows:

- Dimensions of Primary Availability:
  - Geologic (does the mineral resource exist?)
  - Technical (can we extract and process it?)
  - Environmental and social (can we produce it in environmentally and socially accepted ways?)

- o Political (how do governments influence availability through their policies and actions?)
- o Economic (can we produce it at a cost users are willing or able to pay?)
- Dimensions of Secondary Availability:
  - o Technical (can we recover and reprocess it from old scrap?)
  - o Environmental and social (can we recover and reprocess it in environmentally and socially acceptable ways?)
  - o Political (can we assign high political importance to the recovery and reuse of old scrap?)
  - o Economic (can we recover and reprocess it at a cost users are willing or able to pay?)

Two years after this initial criticality concept was formulated, it was utilized (with some revisions) to analyze materials criticality for the European Commission and the US Department of Energy. Since that time, a number of different approaches to critical materials assessment and responses have been devised. Topics that have been emphasized have addressed the importance of scale (national, regional, global), the economic aspects of supply shortages, the treatment of technology in the assessment, or changes in the environmental implications of material acquisition. Regardless of the precise methodology employed or the time scale addressed, the results have tended to focus on materials with low abundance in Earth's crust, those available largely as byproducts, and (for other than global assessments) those with little or no availability in domestic mineral deposits.

16.2 Criticality Methodology

As a representative example of criticality methodology, we illustrate and discuss a comprehensive approach developed at Yale University (Graedel *et al.*, 2012; see Further Reading). The basic concept was to adopt the two axes suggested by the National Academies committee and to add a third axis to address issues related to environmental impacts, where each axis consists of multiple components and indicators. The goal was to develop an approach that could be used at corporate, national, and global levels. The final Yale methodology is shown in Figure 16.2, consisting of a Supply Risk

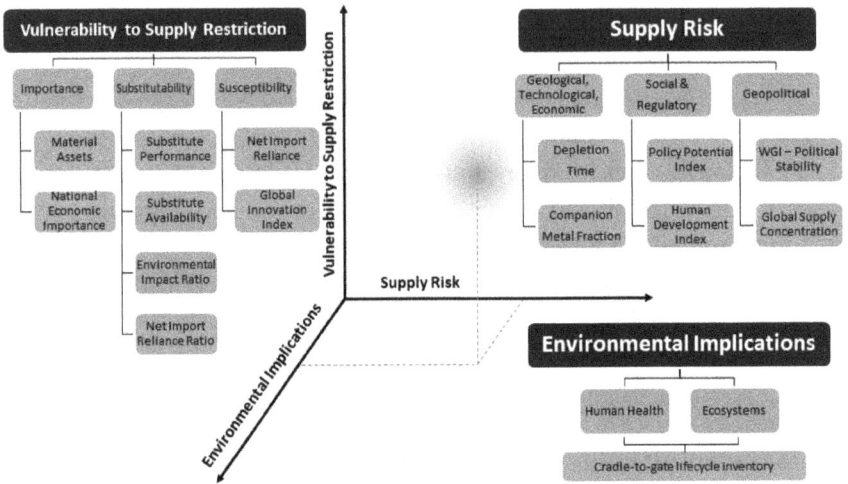

Figure 16.2.   The Yale University criticality analysis methodology across axes of Supply Risk, Vulnerability to Supply Restriction, and Environmental Implications.

axis (SR, upper right), a Vulnerability to Supply Restriction axis (VSR, upper left), and an Environmental Implications axis (EI, lower right).

An example of the results from applying this methodology is given in Figure 16.3, where criticality at the US level (top diagram) and the global level (bottom diagram) is shown for the five elements commonly found in iron mineral deposits. The estimated uncertainties around each axis arise from a Monte Carlo analysis, the results of which form an "uncertainty cloud" for each metal as indicated by the breadth of the plotted points on each axis. For niobium (Nb, shown in green) the uncertainty level is greatest for supply risk, while vulnerability is highest for manganese (Mn, shown in purple). At the global level the supply risk is generally lower because including all mineral deposits around the world provides lower risk, but the vulnerability to supply restriction ratings shown here are roughly equivalent to the US evaluations.

16.3 Alternative Determinations of Material Criticality

The methodology of the European Commission (2017) employs somewhat different metrics distinguished by the imposition of firm lines that

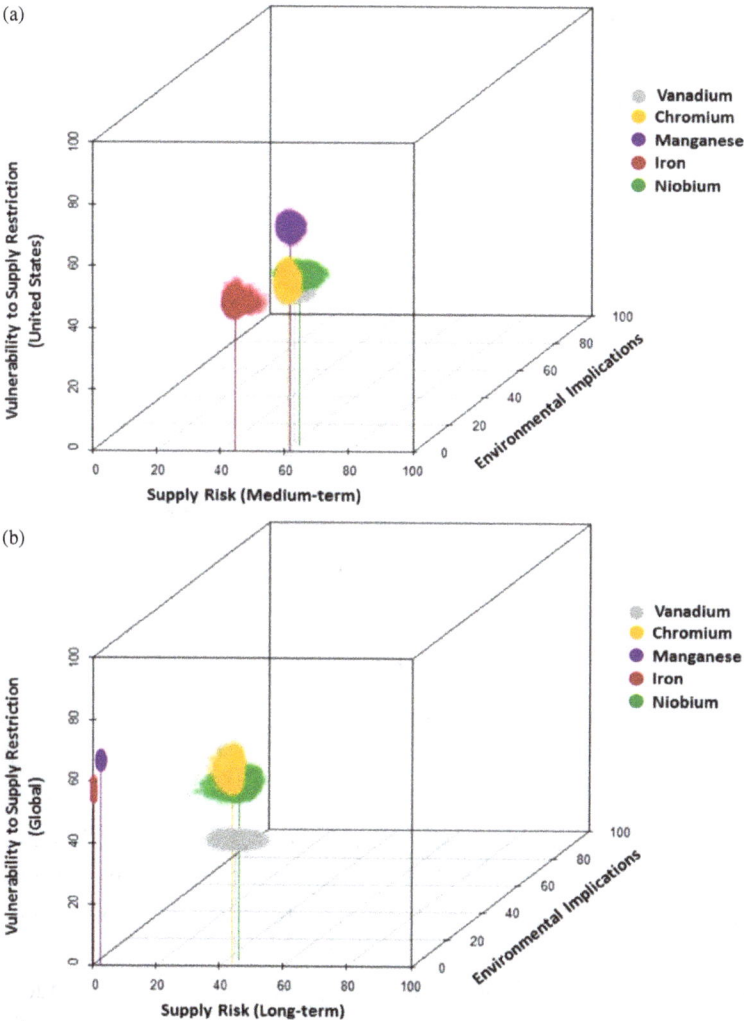

Figure 16.3. Locations of the geological iron family of elements in criticality space (Nuss *et al.*, 2014): Top: national level for the United States, 2008. The highest level of criticality is at (100,100,100) (back right top); Bottom: global level, 2008.

define criticality boundaries for both the axes of Supply Risk and the Economic Importance to the EU (Figure 16.4).

Determinations of material criticality have major importance in the policy realm because country and regional governments are very interested in whether supply/demand imbalances might have impact upon their

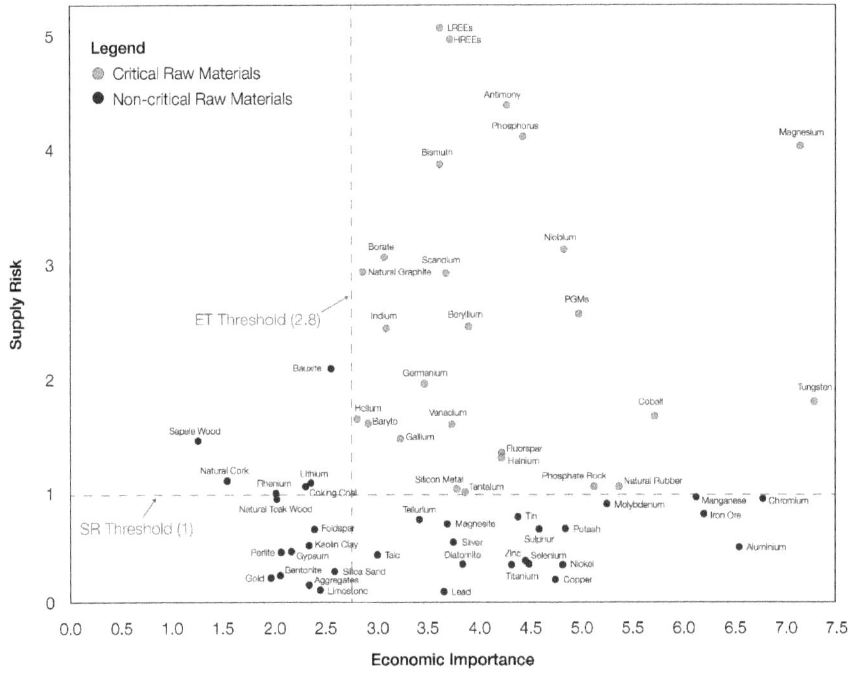

Figure 16.4. Elements deemed critical by the European Union using its methodology, 2017 (adapted from European Commission, 2017).

economies. As a result, there are a number of "criticality lists" that guide governmental policy, including Australia, Canada, the European Commission, Japan, and the United States. Together, the surveys choose a total of sixty-three elements — not too far from selecting all the elements! To be optimally useful, it would seem that criticality methodologies should target the extreme end of the criticality spectrum, perhaps fifteen or twenty elements at most.

16.4 Addressing the Challenges of Critical Materials

The properties and benefits of different classes of materials in design and use vary widely. Depending on the material characteristics that a particular use requires, materials are selected based on the degree to which those properties are met. Material criticality, however, places an additional constraint on the product designer. If a material that is desired is designated as critical by governmental agencies, its supply over the immediate or longer term may be uncertain, and political considerations may also play a role depending on the stability or cooperation of the principal producing countries. These considerations may lead the product designer to seriously consider materials with somewhat inferior mechanical or chemical properties but with much greater security of supply.

From a political perspective, a short-term response to a determination of criticality for a specific material is to establish a stockpile to compensate for scarcity at a particular point in time. Over the longer term, governments can consider increasing supply by enabling new mining projects to proceed under careful supervision. Another alternative is to embrace enhanced recycling, which can be helpful for elements easily recovered and reused (such as platinum-group metals from motor vehicle catalytic converters). In other cases, such as the germanium used to optimize infrared radiation transmission in fiber-optic cables, recovery can be problematic. If concern for germanium supply is high enough, therefore, it may be necessary to redesign the cables in order to minimize germanium use in that application. Similar considerations apply to many other of the materials identified in Figure 16.5.

Further Reading

Australian Government, Australian Critical Minerals Prospectus, 2020. Government of Canada, Canada's Critical Minerals List 2021.

European Commission, *Study on the Review of the List of Critical Raw Materials: Criticality Assessments*, ISBN 978-92-79-47937-3, Brussels, 2017.

Graedel, T.E. *et al.*, Methodology of metal criticality determination, *Environmental Science & Technology*, 46, 1063–1070, 2012.

Graedel, T. E. *et al.*, Criticality of metals and metalloids. *Proceedings of the National Academy of Sciences*, 112(14), 4257–4262, 2015.

McCullough, E., and N.T. Nassar, Assessment of critical minerals: Updated application of an early-warning screening methodology, *Mineral Economics*, 30, 257–272, 2017.

Nakano J., The Geopolitics of Critical Minerals Supply Chains, Center for Strategic & International Studies, 2021. (This is the Japanese criticality assessment.)

Nuss, P., E.M. Harper, N.T. Nassar, B.K. Reck, and T.E. Graedel, Criticality of iron and its principal alloying elements, *Environmental Science & Technology*, *48*(7), 4171–4177, 2014.

US Department of the Interior, *Final List of Critical Materials 2018*, Federal Register, 23295–23296, 2018.

US National Research Council, *Minerals, Critical Minerals, and the U.S. Economy*, Committee on Critical Mineral Impacts on the U.S. Economy, The National Academies Press, Washington, DC., 2008.

Looking to the Future

Chapter 17

The Status of Resource Supply and Demand

Chapter Concepts
• Growing populations and increasing per capita incomes worldwide are resulting in rapidly increasing demands for resources.
• Supplies of some important resources appear unlikely to meet demand by mid-century or end-of-century.
• In the short term, innovative technology may be sufficient to achieve supply-demand balance; in the longer term, constraints on demand may be required.

17.1 How Much Resource Supply is Needed?

The anthropogenic demand for resources is generated by human needs and wants and provides the incentive for acquiring and processing the resources necessary to deliver various services. Consumer demand is generally not for basic resources such as minerals or fibers, but for the variety of industrial products that can be produced from those resources. The links between supply and demand are suggested by the dashed arrows from Industrial Products to Demand segments in Figure 17.1. In most cases, a failure by any of the Industrial Product activities to provide the desired flows to the Demand Segments will result in significant

187

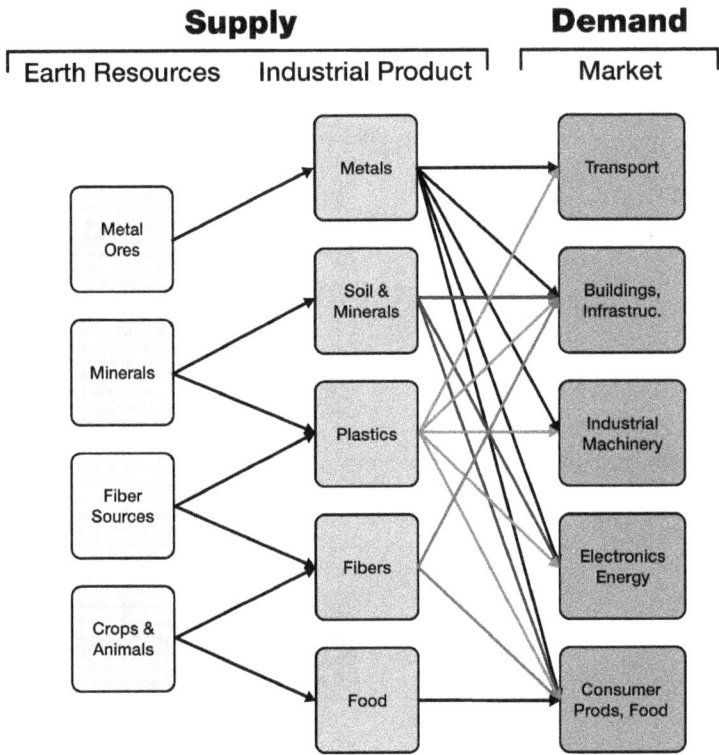

Figure 17.1.   Material flows for the assessment of resource supply and demand.

societal disruption. And, as can be seen, most of the Industrial Product activities serve most of the Demand segments. In fact, the challenge is far greater than the previous statement implies. As chronicled in earlier chapters, the material resources employed in modern technology include some 60 different metals, more than a score of soil nutrients and other mineral additives, perhaps a dozen major plastics, and at least half a dozen fibers. In fact, a survey of the number of elements in modern automobiles lists more than 70. A supply shortage of *any* of these resources or industrial products has the potential to cause disruption and/or hardship. Such supply/demand mismatches may be short-lived and able to be surmounted with effort and expense, but in other cases mismatches may be long-lasting or perhaps permanent. The challenge is for us to look in detail at cases that appear to be in the latter category and to attempt to determine whether long-term mismatches indeed seem likely. If so, what can industrial ecology reveal about the possible response options?

17.2 Scenarios of Resource Supply and Demand

A classic way to imagine the future is to look to the past and present for evidence of change. From the perspective of materials of various kinds, an authoritative compilation of material flows over the period 1970–2010 clearly demonstrates a sharp increase in materials extraction over that period (Figure 17.2). Even more significantly, the rate of increase grew over the 2000–2010 decade, driven by increases in population, affluence, and urbanization around the world (recall the IPAT equation from Chapter 1). Should that trend continue, it is likely to become increasingly challenging for supply to meet demand across the global economy. The sections below explore aspects of supply potential until mid-century for the major groups of materials.

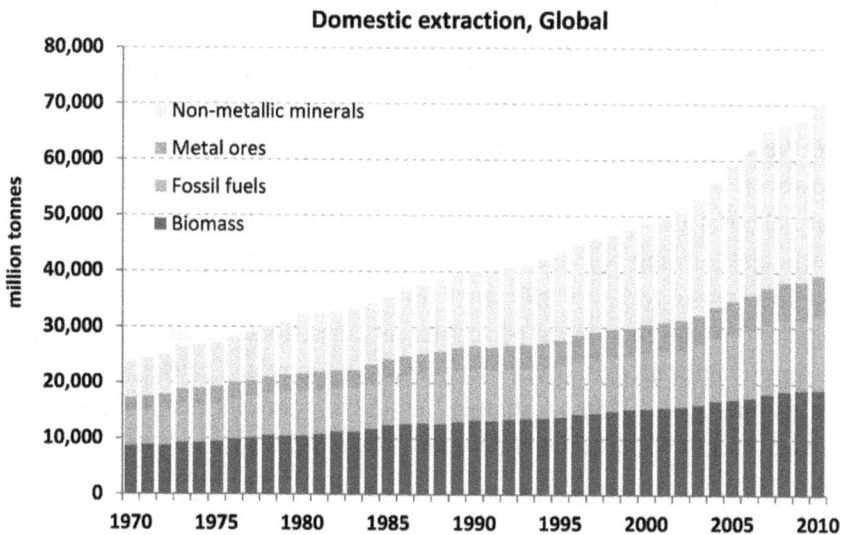

Figure 17.2. The global extraction of four major materials categories, 1970–2010 (Schandl *et al.*, 2018).

17.2.1 *Non-metallic minerals*

Non-metallic minerals are the largest category of domestic extraction globally (as clearly shown in Figure 17.2), generally as a consequence of the use of sand, aggregate, and gravel for building construction and road-building. In principle, at least, the amounts of aggregate and gravel appear unlimited, although sand supplies may become problematic and energy for extraction and processing will be needed, as will material transport from source to point of use.

Of more interest in the present discussion are non-metallic minerals whose properties are unique or unusual and which are significant for society. The obvious minerals in that context are those that enable productive agriculture: nitrogen, potassium, and phosphorous. Nitrogen availability is not at issue, as the Haber–Bosch process extracts it from the atmosphere and converts it to ammonium nitrate. Potash availability also appears not to be a serious issue for the 21st century, as analyses indicate a reserves to production ratio of about 500. ("Reserves" is the total amount of a resource that can be mined with today's technology at today's market prices.)

Phosphorus presents a more complex story. It is derived from phosphate rock that is processed to generate ammonium phosphate fertilizer, and current reserves of phosphate rock (predominantly in Morocco and the Western Sahara) could satisfy present demand for several centuries. However, the phosphorus in fertilizer is dispersed in a number of ways, only about 15% being ultimately consumed in food. The rest is lost to inefficient farming practices, lack of recovery from livestock manure, and low food waste recovery efficiency. Furthermore, given increasing global populations and increasing per-capita incomes, food demand is expected to significantly increase over time. Unless capture and reuse of phosphorus in the food chain is sharply increased, phosphorus demand may exceed supply within several decades.

17.2.2 *Fibers*

Forest products have traditionally been used in the construction industry and for making a wide variety of paper products. In recent times this product spectrum has been diversified: it now includes different types of engineered wood products (cross-laminated timber, for example), as well as forest-based synthetic fibers. With cotton-producing land having high water requirements and often being purchased for more lucrative uses, and the realization that synthetic fiber production and subsequent loss as microplastics is often environmentally harmful, wood or bamboo-based fabrics and other new products are finding market opportunities.

Forest-based manufacturing depends, of course, on the continued existence and competent management of natural fiber sources. Surveys suggest that sufficient land is available and that annual growth in forests is likely adequate to support continued sustainable extraction and production, even as new uses proliferate. The loss of carbon storage due to extensive tree harvesting may, however, complicate the drive to minimize climate impacts (see Peng *et al.*, 2023 in Further Reading).

17.2.3 *Metal ores*

The future prospects for metal ores have been researched much more extensively than for fibers and non-metallic minerals, so it is therefore appropriate to consider metal ore status in some detail. We begin by providing scenario results for prospective demand for seven major metals to 2050 (Figure 17.3). The details are discussed by Elshkaki *et al.* (2018; see Further Reading), but the central message is easy to discern. The diagram shows derived scenario demands as colored wedges, with the vertical axis on the plot expressed in fractions of the "reserves to mid-century (RMC)": the estimated geological supply available to meet the prospective scenario demands. RMC values are only roughly estimated, but the results suggest that for zinc and perhaps for copper, lead, and nickel the existing geological resources may not be sufficient to meet demand.

Another mining result is relevant in this context: most new copper deposits discovered during the past 20 years or so are modest in size, much more so than in earlier epochs, and they tend to be of lower ore grade. An explanation for the ore grade decreases over time is inspired by a diagram originally proposed in 1976 by Brian Skinner, a Yale University geologist (Figure 17.4). Skinner's concept, supported by extensive historic data for moderate to high ore grade mining, was that metals are deposited

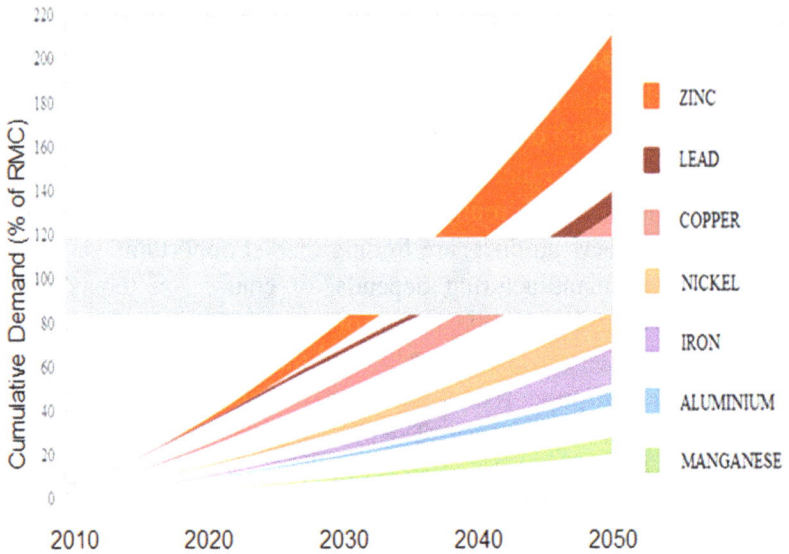

Figure 17.3.   Cumulative global demand for seven major metals, 2020–2050, compared to Resources to Mid-Century (the roughly estimated potential global resource production to 2050, indicated by the gray region). The calculated demand for four different scenarios for a given metal are contained within its colored wedge (Elshkaki *et al.*, 2018).

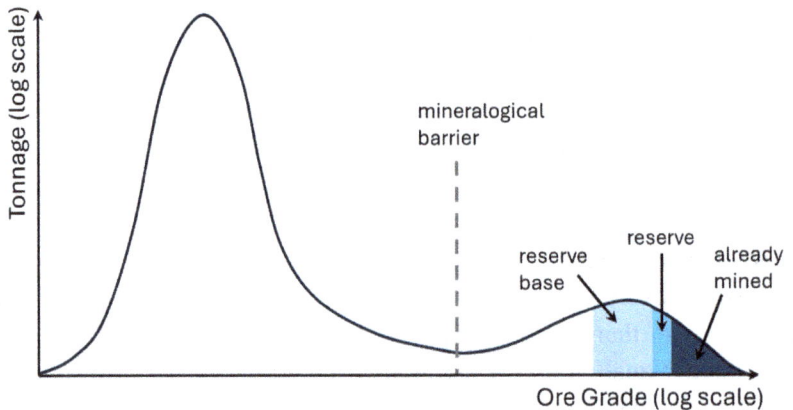

Figure 17.4.   Skinner diagram for copper resources in Earth: already mined, reserves, reserve base, and in common rocks (adapted from Skinner, 1965).

Table 17.1. Estimated ultimately available resources from terrestrial and seafloor repositories (Hein *et al.*, 2013).

Element	Terrestrial resources (Tg)	Seafloor resources (Tg)
Lithium	1.6	1.4
Titanium	899	67
Vanadium	38	9.4
Manganese	5200	5990
Cobalt	13	44
Nickel	150	270
Copper	1000+	226
Arsenic	1.6	1.4
Yttrium	0.5	2
Niobium	3.0	0.46
Molybdenum	19	12
Tellurium	0.05	0.08
Tungsten	6.3	1.3

in rock in two different ways: in a small quantity of high ore grade deposits and in a dispersed molecular fashion in common rocks. Between the two humps on the diagram is a barrier imposed by the processes of mineralogy that occasionally provide favorable circumstances that concentrate molecules into ores. The richest (right-hand) portion of that higher-grade peak has by now been largely mined out, and miners are now working on the Reserve — larger in quantity but poorer in grade. Still larger in quantity but even poorer in grade is the Reserve Base. Beyond that stage is the mineralogical barrier. Extensive evidence to validate Skinner's vision does not exist, as it would require extensive mining of extremely low-grade ores of negligible value. Nonetheless, Skinner's picture is generally accepted by the mineralogical community. If so, future mining is unlikely to locate an abundance of new rich ore deposits.

17.2.4 *Energy considerations*

Fossil fuels are the final major material resource category in Figure 17.2. The world is in the midst of a widespread transition away from fossil fuels and toward electrification and renewable energy. The International Energy Agency projects that global demand for fossil fuels will peak around 2030 and decrease sharply afterwards.

A final energy consideration in the future of resources availability is that extraction and processing of resources can be very energy-intensive, especially as resources become more diffuse and harder to obtain. For metals, for example, as ore grades decrease the energy requirements form essentially a straight line on an energy use-ore grade plot. Metal supplies for future generations thus require (1) locating rich ore deposits, (2) receiving approval for mining and pre-processing, (3) acquiring sufficient non-polluting energy sources, and (4) final processing, purification, and marketing. This list of requirements presents a high bar for metal supply in a period in which every indication is that demand will continue to increase steadily over time.

17.3 The Potential for Curbing Demand

As discussed in Chapter 2, all people wish to satisfy a reasonable set of basic needs: food, housing, safety, heat, and light. After that, they value aspects of a more interesting, valuable, and rewarding life: transportation, adequate medical care, education, and so forth. Our interest is typically not in *how* but *whether* those needs and wants are met, because what we actually need are the services provided by technology, not the technology itself. And in many cases, we cannot see or do not know what materials are being used anyway. For example, building owners might be willing to have engineered wood used in place of concrete and steel in buildings so long as the buildings meet appropriate performance standards and serve their purposes for the occupants. Both materials provide structural stability, but engineered wood is far preferable environmentally. The functions that materials provide can often be fulfilled by multiple options, leaving it to engineers to choose based on cost, performance, and (one hopes) environmental attributes. But in some cases substitution is much more challenging. For a dozen different metals used in modern society, there are essentially no practical alternatives (see Graedel *et al.*, 2015 in Further Reading).

These and other approaches for minimizing the use of materials while maintaining or improving performance are collectively called "material efficiency", Such approaches can minimize product weight by as much as 30% and reduce scrap metal generation in manufacture by up to half. Initiatives such as these hold out the possibility of alleviating materials-related impacts even while populations are rising.

17.4 Weighing the Options

There is general agreement that a satisfactory lifestyle requires at minimum a few basic needs (see Chapter 2): food, drink, warmth, and rest. From a physical perspective this list implies food production, clean water supplies, adequate housing, and personal safety. When these needs are satisfied, personal interactions are next on the list: companionship, mating, social groups, and the like; these are enabled by transportation, communications, and stable societies. From an industrial ecology perspective, most of these needs are met to greater or lesser degrees by the use of materials.

A growing global population implies a concomitant increase in food demand. The availability of productive agricultural land thus has a high priority, but growing more food is only part of the needed response — the food needs to be transported to the people who need it. In a teleconnected world this realization implies large-scale food production and global and local shipping, and thus trucks, merchant ships, distribution centers, and a variety of related infrastructure.

The need for shelter in turn requires materials for housing, procured by activities such as logging, mining, the production of industrial products such as beams and fasteners, intermediate product transport, and assembly equipment. Clean water implies water collection, treatment, transport, and delivery, and thus a variety of pumps, tubes, and pipes. Clearly all of these activities and products require energy, and lots of it. And, as global climate continues to evolve, the energy needed for increased cooling will also rise considerably.

Human companionship, urbanism, and well-being suggest substantial and increasing demand for connected communities, travel over moderate and longer ranges, and a host of life-improving technologies. Therefore, meeting human needs, even imperfectly, will require access to very large quantities of services if humanity is to possess even a modest level of its needs and wants in the future. How we supply these services in material terms is the big question. We are in the middle of a great transition, moving away from our existing fossil and non-renewable industrial system and toward renewable forms of energy, toward bio-based chemicals, toward greater circularity, and toward greater environmental consciousness. Whether the pace of environmental innovation and investment can balance the increased demand from rising population and affluence is a major question for the 21st century to attempt to answer.

Further Reading

Ali, S. *et al.*, Mineral supply for sustainable development requires resource governance, *Nature*, *543*, 367–372, 2017.

Allwood, J.M., and J.M. Cullen, *Sustainable Materials — With Both Eyes Open*, Cambridge, UK: UIT Cambridge, 2012.

Bhuwalka, K. *et al.*, Characterizing the changes in material use due to vehicle electrification, *Environmental Science & Technology*, *55*, 10097–10107, 2021.

Elshkaki, A., T.E. Graedel, L. Ciacci, and B.K. Reck, Resource demand scenarios for the major metals, *Environmental Science & Technology*, *52*, 2491–2497, 2018.

Graedel, T.E., E.M. Harper, N.T. Nassar, and B.K Reck, On the materials basis of modern society, *Proceedings of the National Academy of Sciences*, *112*(20), 6295–6300, 2015.

Peng, L., T.D. Searchinger, J. Zionts, and R. Waite, The carbon costs of global wood harvests, *Nature*, *620*, 110–115, 2023.

Schandl, H. *et al.*, Global material flows and resource productivity: Forty years of evidence, *Journal of Industrial Ecology*, *22*(4), 827–838, 2018.

Skinner, B.J., A second iron age ahead? *American Scientist*, *64*(3), 258–269, 1976.

Vaccari, D.A., S.M. Powers, and X. Liu, Demand-driven model for global phosphate rock suggests paths for phosphorus sustainability, *Environmental Science & Technology*, *53*, 10417–10425, 2019.

Withers, P.J.A. *et al.*, Towards resolving the phosphorus chaos created by food systems, *Ambio*, *49*, 1076–1089, 2020.

Chapter 18

The Circular Economy

Chapter Concepts

- A circular economy is one in which reuse, remanufacturing, and recycling are strongly favored over disposal.
- Factors such as elemental complexity in products, designs that are impediments to efficient recycling, and basic thermodynamics make a comprehensive circular economy impossible.
- Nonetheless, a semi-circular global economy has the potential to provide major benefits for Earth's people and for the planet itself over time.

18.1 The Circular Economy Concept

The concept of the circular economy is of an economic system that aims to achieve the continuous use of material resources, made possible by activities such as remanufacturing and recycling. The Ellen MacArthur Foundation of the United Kingdom, an advocacy organization for the circular economy, promotes a three-pronged effort:

- Keep products and materials in use.
- Regenerate natural systems so as to maximize renewable resource use.
- Design out waste and pollution.

A complete realization of that vision would eliminate waste and thereby reduce the demand for primary materials. The vision of a sustainable Earth system through a circular economy is closely aligned with earlier sustainable design frameworks such as those described in Chapter 13.

The basic idea of the circular economy is to transform our material society from the traditional linear material approach ("dig it up, use it, dispose of it") to one resembling that of Figure 18.1, in which materials retained in the inner circles require less energy and fewer or no new resources to reuse than would be needed for alternative actions in the outer circles. The idea is inherently attractive and has led to circular economy goals, policies, and metrics being taken up by numerous companies and institutions such as the World Economic Forum and the European Commission. An important caveat in all of this is that so long as material demand continues to grow, the processes of reuse, refurbishment, and recycling can offset only a portion of demand, as explained in Chapter 7.

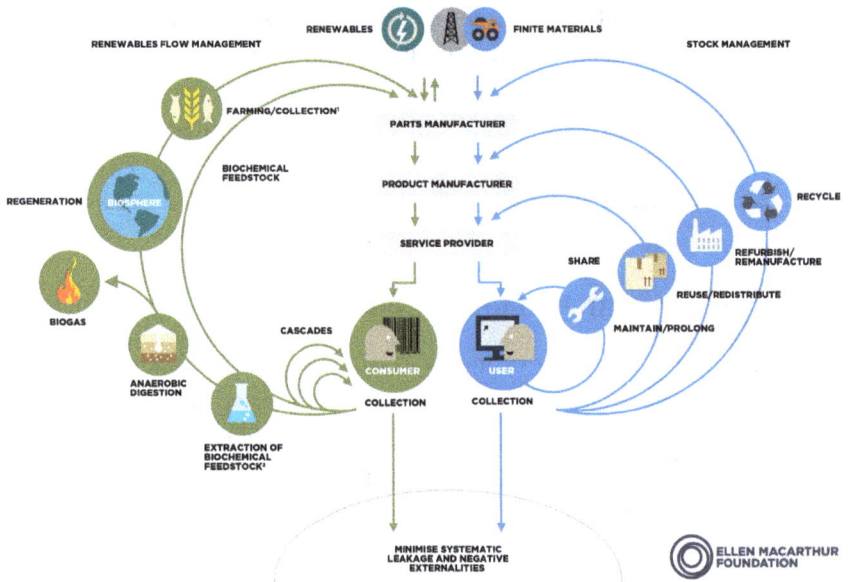

Figure 18.1. A conceptual picture of the circular economy.

Notes: [1] Hunting and fishing. [2] Can take both post-harvest and post-consumer waste as an input.

Source: Ellen MacArthur Foundation *Circular economy systems diagram* (February 2019).

www.ellenmacarthurfoundation.org; drawing based on Braungart & McDonough, Cradle to Cradle (C2C).

Constraining demand is thus a prerequisite for a totally circular economy, no matter how efficient our recycling system might be.

The larger challenge, however, is to determine the degree to which such a transition from the present approach might be possible and, more importantly, might be desirable from technological, environmental, economic, social, and political perspectives. If comprehensive recovery and reuse would require problematic environmental tradeoffs (massive energy use and greenhouse gas emissions, for example), the circular economy, while inherently attractive, may not be environmentally beneficial and should not be an end in itself.

18.2 Materials, Products, and the Circular Economy

To present the scope of the challenge of addressing circularity, it is instructive to discuss a few sample topics that illustrate some of the challenges and nuances of a circular economy that are likely to involve design engineers, consumers, policymakers, and other stakeholders. These and other topics are explored in more detail in Further Reading.

18.2.1 *Metals*

Unlike some other materials, metals are, in principle, capable of being reused indefinitely if they can be recovered with high efficiency and at sufficiently high purity. This is fairly close to the current situation with a number of widely used metals such as copper, in which much of the use is in highly pure forms as an electrical conductor. In many cases, however, metals are used in alloy forms in which mixtures of a few to more than a dozen elements are employed. Many of the alloy metals are "companion metals" that are used largely in trace amounts in complex products. Recovery and reuse of these companion metals depends on the technological potential for separation during reprocessing, which generally involves the treatment of the alloy metals in one or more smelters. As discussed in Chapter 7, the thermodynamics of the smelters and the mixtures of elements typically result in the recovery of some of the elements and the loss of others to slag (furnace waste) or vapor. Partly because recovery of minor alloy constituents has not traditionally been a focus of recycling technology, end-of-life recycling rates of most of these "companion metals" are quite low (see Figure 7.4(b)).

On one hand, the energy required (and thus the climate impacts of energy use) would seem to discourage recovery of small fractions of minor metals from alloys or other mixtures. However, recovering those metals from products is generally less energy intensive than mining them, partly because their concentrations in modern technological products are often higher than they are in mineral ores. Depending, therefore, on the degree to which these metals are being utilized in modern technology, the expenditure of significant amounts of energy to recover and reuse them may be a sensible choice given the alternative of procuring them by mining and processing ores.

18.2.2 *Plastics*

The benefits of modern plastics are enormous and ubiquitous. They preserve food, provide inexpensive household and industrial products, and are major factors in modern medicine. However, at least 80% of all plastics that have ever been made have been discarded — a big loss of resources and a major environmental problem. A key reason for this poor circularity performance is failure to collect plastics following use, especially plastics in food packaging where contamination issues may be involved. In addition, the raw material from which plastics are made is fossil fuels, the use of which exacerbates climate change.

A circular economy for plastics would involve their improved separation and purification in a recycling facility, much the same goals as those of the metals industry.

A final challenge is that plastics entering recycling from consumers, health facilities, and other locales are often too contaminated to be recycled directly. Plastics, however, are complex mixtures of polymers, stabilizers, colorants, biocides, antioxidants, ultraviolet absorbers, fillers, and other additives. Cleaning and removal of additives and other contaminants would be required during the recycling process; otherwise, downcycling into less valuable applications may be the only practical option.

Given the scope of harm caused by plastics pollution in streams, rivers, and oceans, and our growing understanding of microplastics contamination

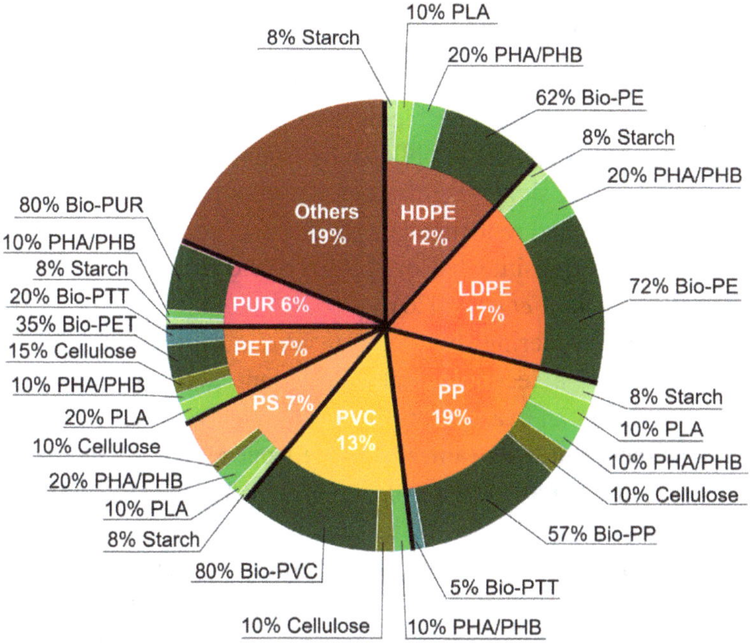

Figure 18.2. The potential for bioplastics to substitute for petroleum-based plastics (Zhao *et al.*, 2020).

at a global scale, plastics are a major focus of circular economy efforts, informed by the industrial ecology tools of MFA and LCA. One push is to redesign plastics, either to be based on biological feedstocks and reduce fossil fuel inputs and emissions, and/or to be bio-degradable and reduce plastic pollution, or both. Figure 18.2 shows that nearly all plastic types currently produced have bio-based substitutes that could be used in many applications.

Another push is to redesign products to use less plastic, more recyclable forms of plastic, or designs that enable recovery of more plastic components. A final push is to improve the efficiency and economics of plastics recycling, where enhanced efforts at collection seem almost certain to be a future focus. Extensive reuse, however, will require behavioral changes in society as well as massive technological development. Some of this will doubtlessly occur, but achieving a semi-circular plastics society appears challenging at best.

18.2.3 *Medical devices*

Medical devices present an interesting case because of the increasing prevalence of single-use disposable devices in the healthcare industry, an approach generally opposite in philosophy from that of the circular economy. In some cases, single-use devices are important for maintaining a sterile environment and reducing the risk of infections, but in many cases, devices can also be disinfected or sterilized and reused without risk to patients (see MacNeill *et al.*, 2020).

It is worth considering the scope of what a truly circular economy would demand of a technologically-proficient multi-product medical device industry. As an example, note that the diversity of elements used by GE Healthcare utilizes (amazingly!) all but eleven of the elements below atomic number 90, nearly a dozen of which are considered critical. This incredible elemental diversity emphasizes the extremely broad scope of modern industrial use of almost all the elements in a variety of settings. Each element's use in medical devices or for medical examinations has a purpose, of course: better imaging of body organs, more precise detection of initial stages of disease, minimization of organism damage during testing, and more. A medical device maker adhering strictly to a circular economy vision will have to develop designs for which the whole suite of valuable materials and components can be recovered (or the product as a whole), cleaned, and safely reused. This would be a major commitment for designers, product manufacturers, and supply chain managers, but would certainly be a part of a comprehensive transition to a circular economy.

18.3 Should *Everything* Be Reused or Recycled?

Reuse and recycling sound as if they are the preferred sustainable option for dealing with the accumulation of discarded products, and in general they are. However, there are instances where routine reuse and recycling may not be the ideal approach. One of the most obvious is where a discarded product contains a historically-employed material that would not now be desired in the economy, particularly materials or assemblages not regarded as hazardous when first used but now of significant concern:

toxic metals such as cadmium in aircraft landing gear, lead in paint, mercury in thermostats, or carcinogenic materials such as polychlorinated biphenyls in transformers. Kral *et al.* (2019) (see Further Reading) suggest that new product designs need to avoid such constituents, and that hazardous older products leaving service or materials dissipated during product use should eventually reach a *final sink*: a repository that either destroys an unwanted substance completely or retains it for a long period so that its reuse can be reconsidered in the future. The process is suggested schematically by Figure 18.3.

Examples of the establishment of final sinks are the deep repositories set up by some countries as the recipients of waste material from nuclear power reactors. Because those repositories store potentially hazardous materials they tend to be controversial, especially to those living nearby. Despite the challenges, however, it is inappropriate to utilize materials known or suspected of toxicity and then to provide no way to deal with them when no longer desired. If these materials are deemed so beneficial to modern technology that society wishes to use them, the challenges to

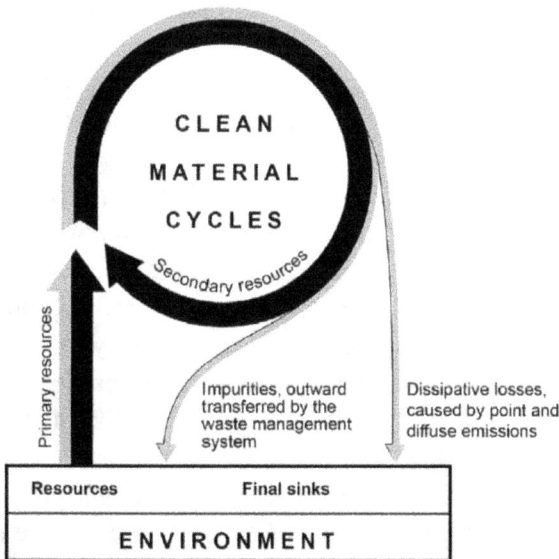

Figure 18.3. Material flows based on a clean cycle strategy (Kral *et al.*, 2019).

doing so need to be recognized and provisions must be made for product life cycles that do not follow a circular economy approach and instead relies on safe waste treatment and disposal.

18.4 Spatial Challenges

One issue not commonly discussed by circular economy advocates is *where* the reuse, remanufacturing, and recycling can or should happen. For simpler materials such as deconstructed building components, markets will need to be relatively close by in order to avoid the expense and challenge of the stockpiling of recovered products. For more complex products, it will be necessary to ensure that they are transported to a facility fully capable of their remanufacture. In a technological world where intricately engineered products are often manufactured in a small number of specialized facilities, sold to users around the world, perhaps later resold or re-leased, and eventually discarded, product complexity and recycling technology cannot be assumed to exist around every corner. As Hagelüken *et al.* (2016) note (see Further Reading), the real challenge is to capture the end-of-life products once they are obsolete but before they become degraded and disassembled, and then to ensure that they are transported to a facility fully capable of their remanufacture. In the case of complex products there will likely be few such facilities in the world, and the challenges of identification, transportation, and economics quickly become daunting.

These locational issues can be illustrated by a simple example, that of nickel metal in Australia, the material cycle for which is shown in Figure 18.4(a). Australia has very large nickel deposits and a vigorous mining industry. As a result, nickel extraction and ore processing are substantial, but the resulting refined metal is largely exported. Much of this nickel goes to be utilized in stainless steel production elsewhere (Australia does not produce stainless steel, an alloy of nickel with about 75 parts iron, 15 parts chromium, and 10 parts nickel). However, imports of stainless steel to Australia (Figure 18.4(b)) might enable a circular economy of nickel if discarded stainless steel were captured for domestic use rather than being mostly exported, but again there are no domestic stainless steel producers to buy the scrap. The message here is that in

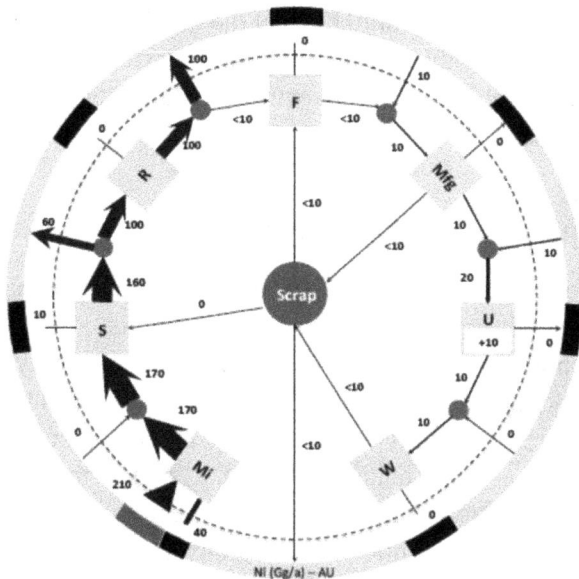

(a)

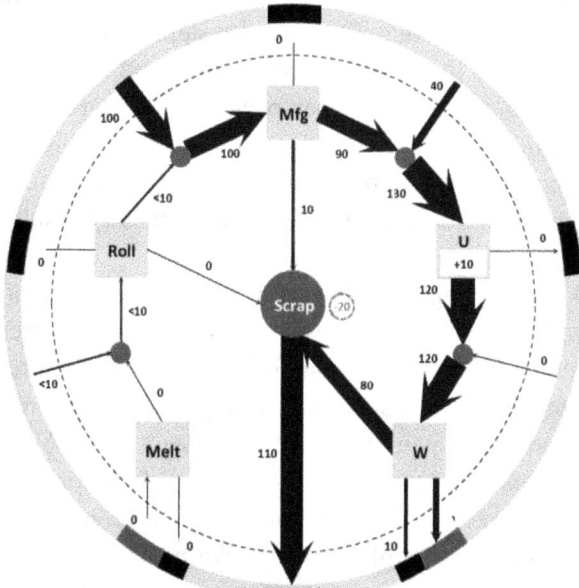

(b)

Figure 18.4. The Australian cycles of nickel (a) and stainless steel (b) in 2010. The units are Gigagrams (thousand metric tons) of metallic equivalent per year (Graedel *et al.*, 2019).

a global economy it is very unlikely that the facilities to enable a circular economy will be available everywhere; rather, extensive ocean shipping and international political and scientific coordination will almost certainly be required.

18.5 Stepping Along the Markov Chain

As we have seen, there are many steps in the life cycles of materials, and losses are virtually certain to occur to some degree at every step. The result of such a sequence can be described mathematically by Markov chain (MC) modeling. In such a system, a unit of material has a specific probability of successfully transitioning to the next stage in the sequence. MCs based on material flow analysis use the distribution of physical flows of materials among economic sectors to populate a transition matrix. If the probabilities of each transition can be estimated, the loss of a material from the economic system over time can be computed. A sample result for nickel flow through 52 geographic regions is shown in Figure 18.5; it reveals that after 50 time steps some 96% of the starting material has been lost, either to the environment or to absorption into iron alloys. From a circularity perspective, the supremely recyclable nickel was found to enter new use over the world only three times. Other studies, for copper and paper, suggest even fewer lifetime reuses. Inexorably, small losses at each life stage ultimately result in material loss rather than eternal reuse.

The nickel recycling story is part of the broader story of thermodynamics, in which the use of energy at each stage of the circularity process will eventually lead to unsustainable levels of resource depletion and generation of waste, especially in a growing economy. This fact should not lead to despair, however. If indeed a completely circular economy is not possible, an extensive and well-functioning semi-circular global economy has the potential to provide major benefits for Earth's people and for the planet itself. We need not completely solve complex challenges to make major strides in the interim in making the situation better. From that perspective, the circular economy goal is one worth aiming for.

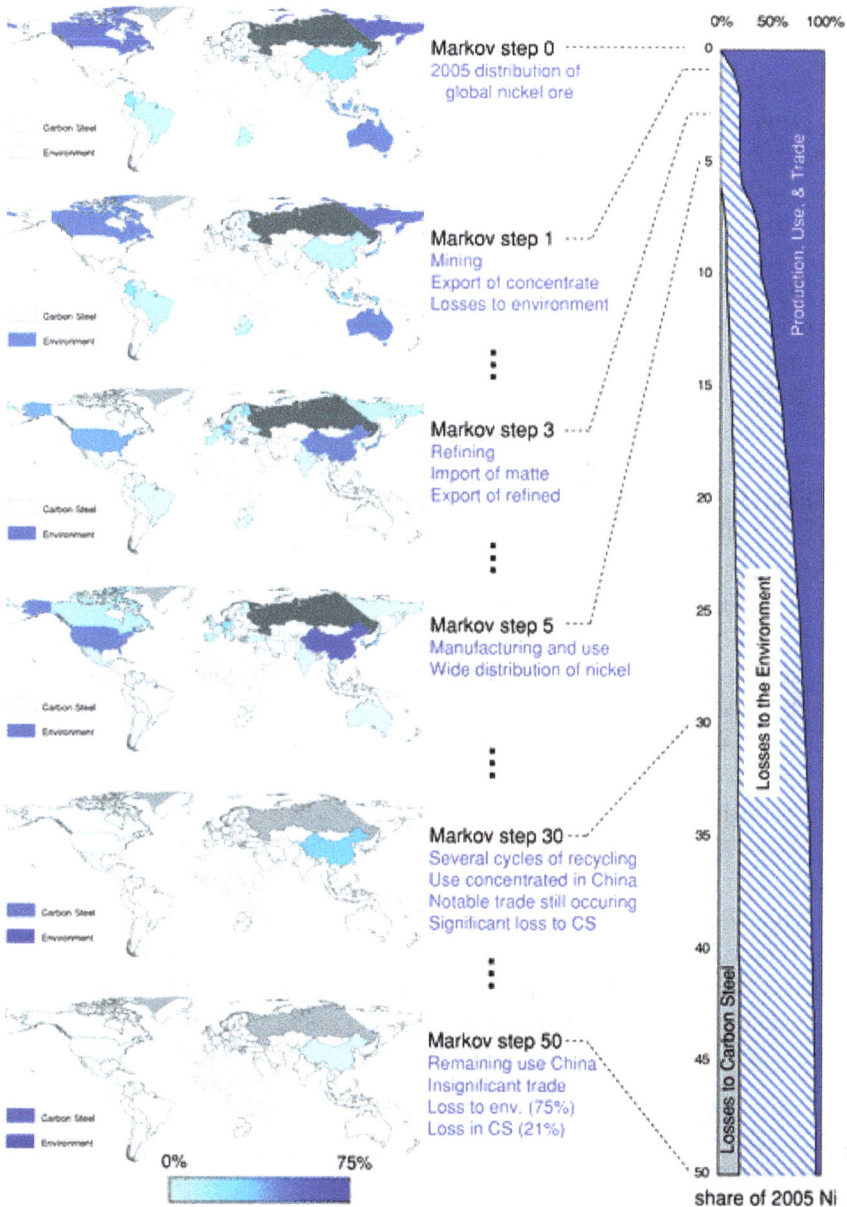

0%   50%   100%

Markov step 0
2005 distribution of
global nickel ore

Markov step 1
Mining
Export of concentrate
Losses to environment

Markov step 3
Refining
Import of matte
Export of refined

Markov step 5
Manufacturing and use
Wide distribution of nickel

Markov step 30
Several cycles of recycling
Use concentrated in China
Notable trade still occuring
Significant loss to CS

Markov step 50
Remaining use China
Insignificant trade
Loss to env. (75%)
Loss in CS (21%)

Production, Use, & Trade

Losses to the Environment

Losses to Carbon Steel

share of 2005 Ni

0%   75%

Figure 18.5. The stepwise global movement of nickel down fifty steps of the Markov chain, in percentage of total 2005 extraction for individual countries, regions, and global sinks (left) and globally aggregated (right) (Eckelman *et al.*, 2012); CS = carbon steel.

Further Reading

Bucknall, D.G., Plastics as a material system in a circular economy, *Philosophical Transactions of the Royal Society A*, *378*, 20190268, 2020.

Eckelman, M.J., B.K. Reck and T.E. Graedel, Exploring the global journey of nickel with Markov models, *Journal of Industrial Ecology*, *16*(3), 334–342, 2012.

Graedel, T.E., B.K. Reck, L. Ciacci, and F. Passarini, On the spatial dimension of the circular economy, *Resources*, *8*, 32, 2019.

Guzzo, D., M.M. Calvalho, R. Balkenende, and J. Mascarenhas, Circular business models in the medical device industry: Paths toward sustainable healthcare, *Resources, Conservation and Recycling*, *160*, 104904, 2020.

Hagelüken, C., J.U. Lee-Shin, A. Carpentier, and C. Heron, The EU circular economy and its relevance to metal recycling, *Recycling*, *1*, 242–253, 2016.

Johnson, J. *et al.*, The energy benefit of stainless steel recycling, *Energy Policy*, *36*, 181–192, 2008.

Korhonen, J., A. Honkasalo, and J. Seppälä, Circular economy: The concept and its limitations, *Ecological Economics*, *143*, 37–46, 2018.

Kral, U., L.S. Morf, D. Vyzinkarova, and P.H. Brunner, Cycles and sinks: Two key elements of a circular economy, *Journal of Material Cycles and Waste Management*, *21*, 1–9, 2019.

MacNeill, A.J. *et al.*, Transforming the medical device industry: Road map to a circular economy, *Health Affairs*, *39*(12), 2088–2097, 2020.

McDonough, W. and M. Braungart, *Cradle-to-Cradle: Remaking the Way We Make Things*. Berkeley, CA: North Point Books, 2002.

Singh, S. *et al.*, Thematic exploration of sectoral and cross-cutting challenges to circular economy implementation, *Clean Technologies and Environmental Policy*, *23*(3), 915–936, 2021.

Van den Burgh, J.C.J.M., Six policy perspectives on the future of a semi-circular economy, *Resources, Conservation and Recycling*, *160*, 104898, 2020.

Chapter 19

Sustainability

Chapter Concepts

- Humanity's current path of resource consumption and environmental degradation appears potentially unsustainable.
- The United Nations Sustainable Development Goals (SDGs) provide a reasonable set of targets for achieving planetary-wide sustainability, equity, and lives worth living. Industrial ecology insights are crucial for achieving many of the SDG goals.
- Sustainability in the 21st century will require a future in which natural and human systems prosper together on an evolving planet.

19.1 Is Humanity's Path Unsustainable?

Societal collapse is an occasional feature of the history of humanity. The classic case is that of Easter Island in the southeastern Pacific Ocean. It is very remote, and was not settled until about 800 A.D. When the Polynesians arrived, they began to cut trees to create farmland and to make canoes. Soon they began to erect the large stone statues for which the island is famous, and trees were used to transport the statues and erect them. Over time, the island's trees were all cut for these purposes.

The lack of trees meant that Easter Island had no firewood, mulch, or canoes. Without the ability to catch dolphins from canoes, and with the depletion of nesting birds, the population came under severe food restrictions, and the island was too remote for help to come. There were no alternatives to a severe and ultimately permanent population collapse.

Easter Island certainly is a special case, but it is not hard to find other cases in which changing conditions or the misuse of technology has forever changed parts of our planet (see Diamond in Further Reading). The discussion of collapse can be generalized by examining the alternative behavioral patterns for complex systems shown in Figure 19.1. The exponential curve (a) traces the path of social progress for some 200 years. This pattern occurs when there are no constraints to growth or when innovation causes apparent limits to recede. Alternatively, the sigmoidal curve (b) is characteristic of the system with fixed constraints in which action is controlled by feedback based on a sense of the distance to the limits.

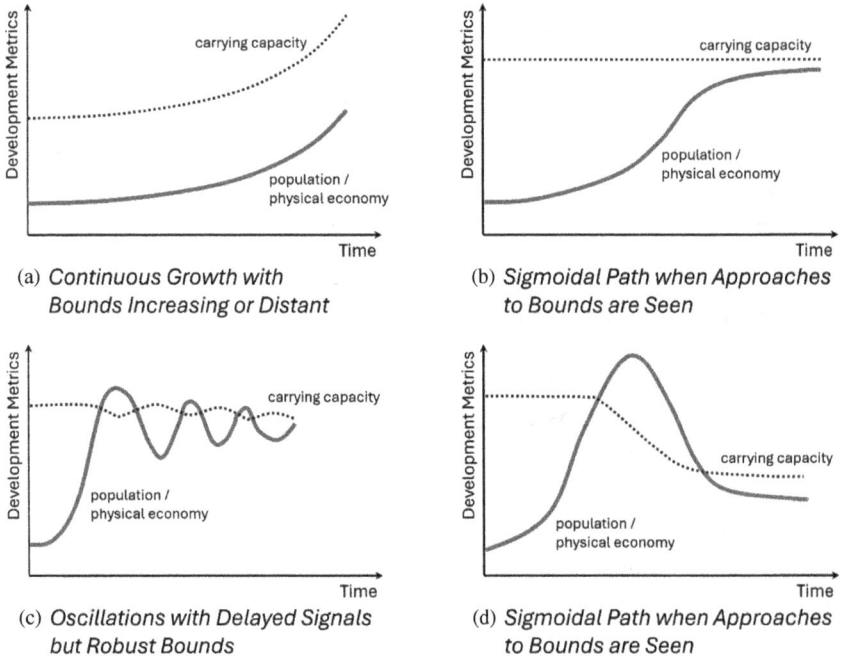

(a) *Continuous Growth with Bounds Increasing or Distant*

(b) *Sigmoidal Path when Approaches to Bounds are Seen*

(c) *Oscillations with Delayed Signals but Robust Bounds*

(d) *Sigmoidal Path when Approaches to Bounds are Seen*

Figure 19.1. Four typical behavior patterns for complex systems (Meadows *et al.*, 1992).

To manage the smooth approach, the system must respond without significant lags and with accurate knowledge of the distance yet to go.

The curve showing oscillatory behavior (c) is typical of systems where feedback mechanisms are inaccurate and responses are slow. At the point that awareness of some limit becomes sufficient to produce action, it is nonetheless too late avoid overstepping the limits and the system continues to move beyond what appears to be some long-run sustainable state. If the stress produced by the overshoot does not completely degrade the system, subsequent corrections can enable the system to oscillate about and approach the limit. Curve (d) depicts initial behavior somewhat like the third curve, but with a critical difference. Here the system is insufficiently robust, corrections are insufficient, and collapse occurs. This is the Easter Island trajectory.

It is important to note that the initial stages of these curves are quite similar. We imagine that we are close to the origin, and further imagine intuitively that we are on the exponential growth pattern. If we find we are not, we must look sustainability in the face and think hard about the robustness and stability of our technological society.

There are possible clues as to what may lie ahead if no significant changes to human resource supply and use are eventually agreed upon. Consider the analogy of the Petri dish — a round flat dish containing a nutrient growth medium onto which are deposited bacteria. As the bacteria grow, they occupy increasing fractions of the available space and consume more and more of the available nutrients. When the bacteria eventually exhaust the food supply, they die. In similar fashion, our modern society consumes industrial food — metals, engineering plastics, fibers, energy on demand. If or when industrial food supplies are no longer sufficient to meet demand, society must somehow reduce demand, switch to other industrial food sources, or disaster awaits.

The Petri dish is useful and defendable in bacteriology given its properties:

- A spatial boundary;
- A defined stock of nutrients;
- A growing population of consumers;
- Stability of external conditions.

If our concern is the human population of the planet, however, the planetary version of the Petri dish has a set of properties that mimics the Petri dish in some aspects but not others:

- An increasing human population;
- An increasing rate of resource demand per individual;
- A potentially decreasing stock of available primary resources;
- A spatial boundary determined by regions suitable for human habitation;
- Rapid changes in a variety of currently enabling conditions, including rising sea-levels, increasing temperatures, and limits to agricultural production.

Most of these properties and their evolution have been discussed in earlier chapters, especially the rapid changes in planetary and societal regimes that have occurred in recent history. Suffice it to say that global change in the Anthropocene is pushing many Earth systems into unknown regimes, and those changes are likely to limit the potential for human societies to adapt given their instinctive tendencies.

19.2 Defining A Sustainability Transition

It is useful in this last chapter to reiterate what we mean by sustainability. Many have tried to formulate succinct definitions, but two stand out. John Ehrenfeld's is conceptual: "*Sustainability* is the possibility that human and other forms of life will flourish on the planet forever". The International Institute of Environment and Development defines *sustainable development* (often used as a synonym for sustainability) as "A development path that can be maintained indefinitely because it is socially desirable, economically viable, and ecologically sustainable". The words have resonance, but offer minimal guidance to engineers, scientists, political leaders, and citizens. Perhaps the most impactful guidance is a famous statement of Mahatma Gandhi: "*There is enough on this planet for everyone's need but not for everyone's greed*". If his philosophy is adopted as a guide for action, perhaps as much out of necessity as conviction, it suggests an uncommon but not undiscussed approach to the future: to move to the creation and maintenance of a modest lifestyle for all rather than a luxuriant lifestyle for a few. The future of the human species may, in fact, depend on whether this transition can be made.

19.3 Approaching Sustainability

As has often been said, "'sustainability" is a term that everyone uses but no one defines. When the issue is addressed in a thoughtful manner, it is generally from the perspective of identifying aspects of behavior that are demonstrably *unsustainable*. This is a useful beginning, but only a beginning. A more useful activity is to measure the attributes of sustainability and then devise ways to move toward a more sustainable state. How is one to approach the challenge of providing sustainability guidance in such a way that it can be implemented? We explore here a few examples of how such guidance might begin to point the way so that the "journey toward sustainability" might continue.

In 2015 the United Nations published a detailed list of seventeen "sustainable development goals [SDGs]" (briefly introduced in Chapter 3). The goals were accompanied by 169 targets that provided additional perspective on actions necessary to achieve the goals. The aim of this initiative was to achieve (or make substantial progress toward) the goals by the year 2030.

As Figure 19.2 illustrates, the goals are quite diverse. Some are hopes for better lives around the world, such as #1: No poverty. Some are environmental in nature, such as #14: Life below water. Others clearly require major doses of resources and technology, such as #7: Affordable and clean energy. However, the case could easily be made that achieving *each* of the seventeen SDGs will require large amounts of resources that are acquired sustainably, used responsibly, and reused extensively. A necessary condition, in fact, is for the world of materials to be linked in detail to the SDGs, for the suite of materials for each SDG to be identified, and for the amount of SDG-related employment of those minerals to be quantified in a defendable manner.

Some SDGs are very closely tied to industrial ecology concerns, such as SDG9-Industry Innovation and Infrastructure and SDG12-Responsible Consumption and Production. But in truth, industrial ecology has an important role to play in designing systems and assessing progress toward

Figure 19.2. The United Nations Sustainable Development Goals (2015).

all of the SDGs, which each involve interlinked human and natural systems and so benefit from an integrated systems perspective.

It is important in this regard not to become trapped into thinking that we must achieve a better world solely by drawing on the technological and social approaches that exist circa 2024. In fact, technology evolves in often meritorious ways and sometimes in transformative ones. Among the transformative examples of early 21st century technology are:

- *Lighting by light-emitting diodes*: LEDs have transformed lighting by providing long-lasting, bright sources of light emanating from small amounts of materials and requiring very modest amounts of energy in order to function.
- *Renewable electricity generation*: Wind turbines and solar PV arrays have become common sights in many parts of the world, supplying carbon-free electricity in place of fossil fuels.
- *Transport by electric vehicles*: EV transport is growing rapidly in both the commercial and consumer markets. These vehicles do not require fossil fuels, and their more limited number of moving parts contributes to enhanced reliability.

- *Construction utilizing "engineered wood"*: Cross-laminated timber (CLT), glue-laminated lumber (glulam), and other evolutionary forest products have made materials such as wood and other fast-growing species such as bamboo increasingly common replacements for heavy traditional steel beams, studs, and other structural components, thereby enabling significant reductions in greenhouse gas emissions while sharply decreasing construction times.

Game-changing technologies like these are generally recognized after the fact, but there is no reason to assume that innovation will not continue to recur. The most useful intellectual areas to pursue will be those that are capable of improving human lives worldwide, especially in the agricultural, medical, and educational sectors. New technologies that meet these criteria should be hunted out and promoted to the world to the greatest degree possible.

At the same time, technology is not a panacea for improving sustainability. If we wait for technological innovation to save the day we may do severe damage in the meantime, and new technologies may have other environmental consequences that we don't foresee. Indeed, there is a school of thought that sees innovation as a driver of *increasing* resource consumption rather than decreasing, for as products and services become cheaper and more accessible, we tend to use more of them, a phenomenon known as the *rebound effect*.

As you have learned, industrial ecology is a systems approach, and its tools and methods explicitly link production and consumption through flows of materials and energy. Technology innovation tends to focus on the supply side of the equation, but making big improvements toward sustainability will require innovation and change on the demand side as well, shifting away from a culture of overconsumption and waste.

19.4 Industrial Ecology for the 21st Century

It is important in the final chapter of this book to recall the framing of the first few chapters: that many of the intellectual perspectives and methodologies of industrial ecology are adapted from those of biological ecology, and that sustainability on a planetary scale will require the efforts of both ecosystem ecologists and industrial ecologists. Now, at the end of this book, we present a final definition of industrial ecology:

> Industrial ecology is the means by which humanity can deliberately approach and maintain sustainability, given continued economic, cultural, and technological evolution. The concept requires that an industrial system be viewed not in isolation from its surrounding systems, but in concert with them. It is a systems view in which one seeks to optimize the total materials cycle from virgin material, to finished material, to component, to product, to obsolete product, and to ultimate disposal.

This call to action retains its focus and relevance in the 21st century, providing insight into how far industrial ecology has come in some 25 years and how much its ideas have become widespread, but also how much is left to do.

Other organisms frequently modify their environments to suit their needs — beavers are an obvious example — but humans' ability to do so is unmatched. As human civilization has expanded through history, our modifications of the planet have been at increasing spatial scales and with increasing impacts. Industrial ecology's principal task is to continue to quantify those changes and to work toward transformative approaches to the use of resources that minimize impacts on the planet and its peoples, while promoting better lives for humans and other organisms, large and small. One cannot imagine a bigger challenge, nor a more important one. We had all better get busy.

Further Reading

Board on Sustainable Development, *Our Common Journey: A Transition Toward Sustainability.* Washington, DC: National Academy Press, 1999.

Diamond, J., *Collapse: How Societies Choose to Fail or Succeed*, revised edition. Penguin, 2011.

Ehrenfeld, J.R., Would industrial ecology exist without sustainability in the background? *Journal of Industrial Ecology*, *11*(1), 73–84, 2007.

Meadows, D.H., D.L. Meadows, and J. Randers, *Beyond the Limits*. White River Junction, VT: Chelsea Green, 1992.

Palmer, M.A. *et al.*, Ecological science and sustainability for the 21st century, *Frontiers in Ecology and the Environment*, *3*(1), 4–11, 2005.

United Nations: *Transforming Our World: The 2030 Agenda for Sustainable Development*, New York, 2015.

United States Geological Survey, *Mineral Commodity Summaries*. Reston, VA, 2022.

Appendix

Units of Measurement in Industrial Ecology

Table A.1. Prefixes for large and small numbers.

Power of ten	Prefix	Symbol
+24	yotta	Y
+21	zetta	Z
+18	exa	E
+15	peta	P
+12	tera	T
+9	giga	G
+6	mega	M
+3	kilo	k
−3	milli	m
−6	micro	μ
−9	nano	n
−12	pico	p
−15	femto	f
−18	atto	a
−21	zepto	z
−24	yocto	y

The basic unit of energy is the joule (= 1×10^7 erg). One will often see the use of the British Thermal Unit (Btu), which is 1.55×10^3 J. For very large energy use, a unit named the *quad* is common; it is shorthand notation for one quadrillion British Thermal Units. Thus, 1 quad = 1×10^{15} Btu = 1.55×10^{18} J = 1.55 exajoules (EJ).

The units of mass in the environmental sciences and in this book are given in the metric system. Since many of the quantities are large, the prefixes given in Table A.1 are common. Hence, we have such figures as 2 Pg$=2\times10^{15}$ g. Where the word tonne is used, it refers to the metric ton $=1\times10^6$ g.

The most common way of expressing the abundance of a gas phase atmospheric species is as a fraction of the number of molecules in a sample of air. The units in common use are *parts per million* (ppm), *parts per billion* (thousand million; ppb), and *parts per trillion* (million million; ppt), all expressed as volume fractions and therefore abbreviated ppmv, ppbv, and pptv to make it clear that one is not speaking of fractions in mass. Any of these units may be called the *volume mixing ratio* or *mole fraction*. Mass mixing ratios can be used as well (hence, ppmm, ppbm, pptm), a common example being that meteorologists use mass mixing ratios for water vapor. Since the pressure of the atmosphere changes with altitude and the partial pressures of all the gaseous constituents in a moving air parcel change in the same proportions, mixing ratios are preserved as long as mixing between air parcels can be neglected.

For constituents present in aqueous solution, as in seawater, the convention is to express concentration in volume units of moles per liter (designated M) or some derivative thereof [one mole (abbreviated mol) is 6.02×10^{23} molecules]. Common concentration expressions in environmental chemistry are millimoles per liter (mM), micromoles per liter (µM), and nanomoles per liter (nM), Sometimes one is concerned with the "combining concentration" of a species rather than the absolute concentration. A combining concentration, termed an *equivalent*, is that concentration which will react with 8 g of oxygen or its equivalent. For example, one mole of hydrogen ions is one equivalent of H^+, but one mole of calcium ions is two equivalents of Ca^{2+}. Combining concentrations have typical units of equivalents, milliequivalents, or microequivalents per liter, abbreviated eq/L, meq/L, and µeq/L. A third approach is to express concentration by weight, as mg/L or ppmw, for example. Concentration by weight can be converted to concentration by volume using the molecular weight as a conversion factor.

Acidity in solution is expressed in pH units, pH being defined as the negative of the logarithm of the hydrogen ion concentration in moles per liter. In aqueous solutions, an acidity of pH = 7 is neutral at 25°C; lower pH values are characteristic of acidic solutions, higher values are characteristic of basic solutions.

Index

www.ingramcontent.com/pod-product-compliance
Lightning Source LLC
Chambersburg PA
CBHW050557190326
41458CB00007B/2073